Grapevine Rootstocks

Selection and Management for
South Australian Vineyards

Nick Dry graduated with a Bachelor of Agricultural Science (Viticulture) from the University of Adelaide in December 2000. He was employed by the Orlando Wyndham Group as viticulturist and spent time based in Padthaway and the Riverland. After spending 7 months travelling ,which included a vintage in the Franken region of Germany and visiting wine regions in Spain and Italy, Nick was appointed to the newly-created position of Rootstock Project Manager with the Phylloxera and Grape Industry Board of South Australia.

Grapevine Rootstocks

*Selection and Management
for South Australian Vineyards*

NICK DRY

PHYLLOXERA AND GRAPE INDUSTRY BOARD OF SOUTH AUSTRALIA

LYTHRVM

First published in 2007 by

Lythrum Press
PO Box 243 Rundle Mall, Adelaide, South Australia 5000
www.lythrumpress.com.au

in association with

Phylloxera and Grape Industry Board of South Australia
46 Nelson Street, Stepney, South Australia 5069
www.phylloxera.com.au

Phylloxera and Grape Industry
Board of South Australia

ISBN 978 1 921013 14 0

Cover design: Stacey Zass, Colorperception, Melbourne
Back cover photograph: Sunridge Nursery, Bakersfield, California (Nick Dry)
Design and typesetting: Michael Deves, Adelaide
Printed and bound by: Hyde Park Press, Adelaide

Contents

Glossary vii
Foreword ix

Background 1
Benefits of Using Rootstocks 2
 Production benefits 2
 Financial benefits 3
Rootstock Characteristics 5
 Vitis species 5
 The characteristics of the common rootstock hybrids 6
 CSIRO (Merbein) rootstocks 19
 Specialty rootstocks 23
Site Factors 24
 Soil properties 24
 Chemical properties 28
 Drought tolerance and water availability 32
 Salinity 33
 Nematodes 34
 Climatic conditions 36
 Fruitset 38
 Vineyard variability 39
Regional Rootstock Recommendations 43
Choosing a Rootstock 66
Managing Rootstocks 72
Further Reading 77
References 77
Appendices 80
 Appendix 1: Rootstock characteristics summary 80
 Appendix 2: Planting statistics 81
 Appendix 3: On-farm rootstock trials 82

Acknowledgements

I would like to thank the PGIBSA board members and in particular the members of the Vine Health Production committee; Dr Jim Hardie (Chairman), Peter Stephens, Robin Nettlebeck and Peter Balnaves for their technical support and feedback during the production of this publication. Special thanks also to Peter Hackworth and Sandy Hathaway for their editorial guidance and many thanks to the other PGIBSA staff for making life in the office so much fun.

To the many the growers whose vineyards I have had access to and to the many growers, viticulturists, nurserymen, researchers and winemakers with whom I have spent countless hours discussing rootstock performance and selection, thank you – without your contribution this publication would never have happened.

Thanks Dad for everything.

And lastly I would like to acknowledge those people of have made significant contributions to rootstock research in Australia; Dr Peter May, Richard Cirami, Dr Mike McCarthy, Phil Nicholas, John Whiting, Dr Rob Walker and Peter Clingeleffer.

Nick Dry

Glossary

Definitions

Rootstocks: are hybrids which incorporate the genes of American *Vitis* species and provide the root portion of the grapevine to which a fruiting vine of any desired variety (such as Shiraz or Chardonnay) can be grafted. The grafting does not produce new varieties since both rootstock and scion retain their characteristics. Most rootstocks in viticulture are used because they confer resistance to the soil-borne pests phylloxera and nematodes.

Ungrafted/own-rooted vines: these terms are used interchangeably throughout the publication and refer to ungrafted vines, i.e. Shiraz, Chardonnay etc, that are planted and growing on their own roots. These vines are susceptible to phylloxera and nematode infestation.

Vigour: in this publication the term 'vigour' is used to describe the total weight of shoots and fruit produced by a vine in a growing season.[1]

Vine balance: when 'the vegetative and fruit load are in equilibrium and consistent with high fruit quality' (Gladstones, 1992).

Site potential: describes the potential vigour (low, moderate or high) that will be conferred to a vine at a given site. Site potential is a function of: soil depth, soil fertility and climate.

Resistance: (from Whiting, 2004) in relation to nematodes and phylloxera, resistance refers to the vine's ability to maintain performance (yield and quality) in the presence of the pest compared with more susceptible varieties. Rootstock resistance to phylloxera and nematodes is mostly by tolerance (see below) but may involve non-preference (vine characteristics that are unattractive to feeding, reproduction and shelter) or antibiosis (vine adversely effects the growth and reproduction of the pest).

Tolerance: the ability of the vine to maintain performance despite supporting a pest population or growing in conditions that would negatively affect the performance of more susceptible varieties.

Susceptibility: the inability to maintain performance under stress caused by soil conditions, the environment, a pest or disease.

---

1. Strictly speaking 'vigour' is defined as the 'rate of shoot growth', whereas the 'total weight of shoots and fruit' is defined as 'capacity'.

Acronyms

CSIRO   Commonwealth Scientific and Industrial Research Organisation

PGIBSA   Phylloxera and Grape Industry Board of South Australia

PIRSA   Department of Primary Industries and Resources of South Australia

SARDI   South Australian Research and Development Institute

Foreword

I am pleased to present *Grapevine Rootstocks: Selection and Management for South Australian Vineyards*.

Grapevine rootstocks have the potential to provide a range of viticultural benefits to wine growers in South Australia. The Phylloxera and Grape Industry Board of South Australia initiated the process of increasing the awareness of rootstocks early this decade and published *A Grower's Guide to Choosing Rootstocks* in 2003. During 2004, Nick Dry was employed to lead a project whose ultimate aim was to address the issues inhibiting widespread rootstock use in South Australia. This project focused on the collection of viticultural and wine quality information from existing commercial grafted vineyards.

This publication is the culmination of the first three years of this project and includes information from a range of other sources including government research agencies (GWRDC, CRCV, PIRSA, SARDI & CSIRO), wine growers, winemakers and nurserymen, and I acknowledge their willingness to assist.

Grapevine Rootstocks: Selection and Management for South Australian Vineyards is designed to support wine growers in making informed decisions on the most appropriate rootstocks for their variety, soil and region.

I acknowledge the efforts of Nick Dry for the coordination and production of this document.

This publication has been collated by the Phylloxera and Grape Industry Board of South Australia on behalf of South Australian growers but I would recommend it to all wine growers in Australia.

Peter Stephens
Chairman
Phylloxera and Grape Industry Board of South Australia

Background

In 2003 the Phylloxera and Grape Industry Board of South Australia published the first edition of *A Growers' Guide to Choosing Rootstocks* after recognising that there was a need to increase awareness amongst growers of the potential viticultural and risk management benefits of planting new vineyards on phylloxera-resistant rootstocks. From 2004–2007 research by PGIBSA has significantly increased the quality and quantity of information on the performance and management of rootstocks in commercial vineyards. This new information has been incorporated into this new publication and will further assist grapegrowers in selecting and managing rootstock varieties for new vineyard plantings and redevelopment projects.

Rootstock performance is specific to local conditions so *Grapevine Rootstocks: Selection and Management for South Australian Vineyards* has a strong focus on the provision of regional information on rootstock performance and selection. A section on 'Managing Rootstocks' has also been included to assist growers to achieve consistent quality outcomes. While phylloxera remains a significant threat to grape-growers in South Australia, there are other more immediate threats to sustainable grape production in the form of reduced water availability and increasing soil and water salinity. In response, this new publication contains an expanded section on the salinity tolerance and water-use efficiency/ drought tolerance of rootstocks.

Viticultural information

The viticultural information contained in this document represents a summary of the most up-to-date information available from Australia and overseas. This publication builds on the information from the earlier *A Growers' Guide to Choosing Rootstocks* and incorporates research conducted by PGIBSA from 2004–2007 with information from a number of other sources including GWRDC-funded projects, government research agencies (PIRSA, SARDI, CSIRO) and anecdotal information from viticul-turists, growers, nurserymen and winemakers.

Disclaimer

While PGIBSA has made every effort to ensure the accuracy of the information contained in this document, it accepts no liability for the use of the information and any consequences that may arise from doing so.

Benefits of Using Rootstocks

Protection against Phylloxera

Phylloxera attacks vine roots in order to feed. As well as direct damage, the damage to the roots caused by feeding leaves the vine open to secondary fungal and bacterial infections. In susceptible vines (e.g. *Vitis vinifera*) this results in the rapid decay of root tissue and eventual vine death. Rootstocks (hybrids of American *Vitis* spp.) are able to resist infection primarily through the formation of a 'protective' corky layer around the feeding site which limits the spread of decay into the vine root. Grafting *Vitis vinifera* to resistant rootstocks is therefore the only effective management strategy for growing grapevines in phylloxera-infested soils.

While rootstock use has increased from 18% to 20% overall in South Australia between 2003 and 2007, many regions and growers still have less than 10% [2] of plantings on rootstocks (see Appendix 2 for the regional breakdown). It is important that regions and growers within regions actively seek to increase rootstock use so that in the event of a phylloxera outbreak they can continue to generate income and fulfil market demands.

Production benefits

Rootstocks should be seen as an extra management tool that growers can use to optimise profitability and sustainability. They can provide the following viticultural/ production benefits:

➤ Nematode resistance (see page 34)
➤ Salinity tolerance (see page 33)
➤ Drought tolerance (see page 32)
➤ Soil acidity tolerance (see page 28)
➤ Influence scion vigour (see page 41)
➤ Influence fruitset (see page 38)
➤ Influence rate of ripening (see page 36)

Long-term sustainability

It is now very clear that growers will have to get by with less water. Linked with declining water availability is the issue of declining water quality and increasing

---

2. As at May 2007

salinity levels in the root-zone. While there are a number of irrigation techniques which reduce water-use such as sub-surface irrigation, regulated deficit irrigation (RDI) and partial root-zone drying (PRD), rootstocks can provide permanent solutions to both declining water availability and declining water quality. As part of their risk management strategy, growers should consider planting a proportion of their vines on a range of drought- and salinity-tolerant rootstocks in a manner that allows for their quality and general performance to be assessed. This will:

➤ Provide growers with insurance in the event that the current situation deteriorates

➤ Assist growers in selecting rootstocks best suited to their site.

For more information on salinity and drought-tolerant rootstocks go to pages 32 and 72. For information on how to establish a rootstock trial on your property go to Appendix 3.

Financial benefits

Increased production efficiency

Scientific and commercial evidence suggest that rootstocks generally increase yield compared with ungrafted vines and where a rootstock is well matched to a site and appropriately managed it may do so without a negative effect on quality. Thus, rootstocks can increase production efficiency (higher returns, for the same inputs) and therefore grower profitability compared with ungrafted vines.

Gawel et al. (2000) conducted a trial in Langhorne Creek that compared the wine spectral and sensory properties of Cabernet Sauvignon grafted to four rootstocks (Ramsey, 110 Richter, Schwarzmann, 5C Teleki) and on its own roots. They found that with higher yields 'the rootstock 5C Teleki in two out of the three years, and Schwarzmann in all three years produced colour and phenolic indices very similar to that of the ungrafted control'. In the final year of the study, wines were made and in the sensory evaluation, despite yielding an extra 6 kg/vine (6.7 t/ha) compared with the ungrafted treatment, the 5C Teleki treatment was described as 'the best of them'.

In another trial, Ewart et al. (1993) studied the effect of rootstock on the composition and quality of Chardonnay in McLaren Vale. This study found that there was no significant difference in quality across the three years of the study between the four rootstocks (Ramsey, 5C Teleki, Schwarzmann, 140 Ruggeri) and the ungrafted treatment despite the rootstocks producing higher yields (between 3.5 kg/vine to 10 kg/vine difference compared with the ungrafted vines).

This response will not occur at all sites or with all rootstocks so a degree of expertise and experimentation is required to identify the rootstock that will lead to greater production efficiency at a particular site. The potential for increased profits should motivate growers to plant small-scale rootstock trials. For information on setting up small-scale rootstock trials see Appendix 3.

Rootstock Characteristics

VITIS SPECIES

All the common rootstock hybrids used in South Australia were bred from the following American Vitis species: *Vitis riparia, Vitis rupestris, Vitis berlandieri* and *Vitis champini*. The rootstocks that result from the crossing of these parent species have characteristics that reflect those of their parents. Understanding the basic characteristics of the parent species can help give growers an insight into the expected performance of a rootstock at a particular site. The following information on the characteristics of the *Vitis* species used for rootstock breeding is taken from Cirami (1999).

Vitis riparia

V. riparia is a trailing or climbing vine. Of all the American *Vitis* species it has the greatest geographical range. It extends from the centre of Canada in the north, to Texas and Louisiana in the south and west to the Rocky Mountains. In its natural environment it grows along the edges of rivers and streams or in shady, moist forests. *V. riparia* prefers cool, moist environments where there is an uninterrupted supply of water. Its roots tend to grow laterally rather than vertically and it is susceptible to drought conditions and performs poorly in sandy or calcareous soils. It has a short vegetative cycle adapted to low light, cool conditions, has some tolerance of 'wet feet' and has very high phylloxera resistance. Cuttings of *V. riparia* root very easily and graft well. Vines grafted to *V. riparia* produce good crops, ripen early and develop a high sugar content.

Vitis rupestris

This species occurs commonly in stony soils in south-western Texas extending northward and eastward to New Mexico, Indiana, Tennessee and Pennsylvania. *V. rupestris* prefers deep, gravelly, rocky soils next to mountain streams. It requires a deep soil and penetrable subsoil. It has moderate lime tolerance (less than *V. berlandieri* and more than *V. riparia*) and a long vegetative cycle. *V. rupestris* roots and grafts easily for nursery production but is sensitive to root-rotting fungi.

Vitis berlandieri

A native of the limestone hills of south-west Texas, this species is found in New

Mexico and north Mexico where it thrives in calcareous, non-fertile soils and dry, hot climates. This species has developed a tolerance to lime soils and drought conditions, and has a long vegetative cycle. *V. berlandieri* is highly resistant to phylloxera.

Vitis champini

Possibly a natural hybrid between *V. rupestris* and *V. candicans*. It is vigorous, resistant to nematodes and tolerates lime soils.

The characteristics of the common rootstock hybrids

The common rootstocks can be divided into four groups based on the parents from which they were bred. The rootstocks within each group have similar viticultural characteristics. Please note that all rootstock use data is from the PGIBSA vineyard register and is current as at May 2007.

V. riparia × *V. rupestris*

These rootstocks impart low to moderate vigour to the scion, hasten ripening but do not tolerate drought conditions. These characteristics make them particularly suited to cool climate viticulture. The varieties include:

➤ 101–14
➤ Schwarzmann
➤ 3309C

➤ **101–14**

101–14 accounts for 6% of total rootstock plantings in South Australia but has been used extensively in France, New Zealand, South Africa and North America.

Characteristics

- Moderate yields in cool and warm climates, relatively high yields in hot climates (Cirami, 1999)
- Low vigour (Galet, 1998)
- Poor performance in drought conditions and high water requirements
- Moderate tolerance of saline conditions (Tee and Burrows, 2004).

6

- Good resistance to root-knot nematode, low resistance to root-lesion and citrus nematode (Whiting, 2004 and Nicol et al., 1999).
- Low-moderate lime tolerance—up to 9% active lime (Galet, 1998)
- May advance ripening, earliest ripening of the group (Galet, 1998)
- May improve fruitset (May, 2004)

✔ *Suitable for:*

101–14 performs best on soils that dry out slowly and have moderate to high water holding capacity, such as moderate to deep loamy sands and fine textured (clay) soils.

It is well suited to and has proven quality outcomes in cool climates because it imparts low vigour to the scion, may advance maturity and improve fruitset. It is worth noting that 101–14 is particularly well suited to Chardonnay, Sauvignon Blanc and Pinot Noir in the Adelaide Hills because of its devigorating effects.

101–14 has been used with success in warm climates (Barossa Valley and McLaren Vale), particularly with Shiraz and Cabernet Sauvignon, but should only be used at sites with adequate and consistent water supplies.

In the Riverland it has consistently produced high berry homogenate colour scores and moderate to high yields; however, considering its high water requirements and the possibility of future water restrictions in this region, it may not be the best long-term choice.

✘ *Not suitable for:*

101–14 has poor drought tolerance so should not be used at sites where water is not available on demand. Avoid sand and sandy-loams sites in warm and hot climates, particularly with white varieties, as late season leaf loss can lead to excessive sun exposure. 101–14 performs poorly on acid soils (Cirami, 1999) and is susceptible to the effects of waterlogging during establishment (Nicholas, web citation).

➤ Schwarzmann

Schwarzmann was used extensively in the late 1980s and early 1990s and prior to 1996 it was the 2nd most widely planted rootstock in South Australia. Since then, cutting sales of Schwarzmann have dropped in favour of 101–14 and the drought-tolerant rootstocks. As of 2006 it was ranked 4th behind Ramsey, 1103 Paulsen and 140 Ruggeri accounting for 8% of total rootstock plantings.

Characteristics

- Low-moderate yields (Cirami, 1999)
- Low-moderate vigour
- High water requirements and poor performance in drought conditions
- Moderately tolerant of saline conditions (Tee and Burrows, 2004)
- Good resistance to root-knot, citrus and dagger nematode, but low resistance to root-lesion nematode (Whiting, 2004 and Nicol et al., 1999)
- High potassium uptake which can negatively affect juice pH (Ewart et al., 1994)
- Moderate lime tolerance—more tolerant than other *V. riparia* × *V. rupestris* hybrids
- May advance ripening (Whiting, 2003)
- May improve fruitset (Cirami, 1999)
- Tolerates waterlogged soils (Whiting, 2003)

✔ *Suitable for:*

Schwarzmann appears to be well adapted to a range of soil types but does best on deep, moist soils. It has a higher tolerance of lime than other *V. riparia* × *V. rupestris* hybrids (Cirami, 1999) and is less susceptible to waterlogged soils than most other rootstocks.

Schwarzmann is suited to and has proven quality outcomes in cool climates because it imparts low-moderate vigour to the scion and can advance maturity and improve fruitset. It is worth noting that Schwarzmann is particularly well suited to Chardonnay, Sauvignon Blanc and Pinot Noir in the Adelaide Hills because of its low-moderate vigour.

Schwarzmann has been used with success in warm climates, particularly with Shiraz and Cabernet Sauvignon; but because it does not tolerate drought conditions it should only be used at sites with adequate and consistent water supplies.

✘ *Not suitable for:*

Its use should be avoided on acidic soils that have not been ameliorated (Whiting, 2003). Also because it has poor drought tolerance it should not be planted in shallow soils or at sites where periods of prolonged water deficit are common. High potassium uptake can negatively affect juice pH and quality.

➤ 3309 Couderc

There are no commercial plantings of 3309C in South Australia but it is one of the most widely planted rootstocks in France, New Zealand and the north-east of the United States.

Characteristics

- Moderate yields (Whiting, 2003)
- Low vigour
- Drought susceptible—more so than 101–14 and Schwarzmann
- Poor performance in saline soils (Tee and Burrows, 2004)
- Low root-knot and citrus nematode resistance, moderate resistance to root-lesion nematode and low-moderate resistance to dagger nematode. (Whiting, 2004 and Nicol et al., 1999)
- Moderate lime tolerance—up to 11% active lime (Galet, 1998)
- Resistant to crown gall, but susceptible to *phytophthora* (Southey, 1992)
- May advance maturity
- May improve fruitset (May, 2004)

✔ *Suitable for:*

3309C has a deep growing, well-branched root system and is best suited to deep, fertile, well-drained cool soils that are well supplied with water.

There is little experience with this variety in Australia, but based on overseas experience it should be a good rootstock for the production of high quality wines in the fertile soils of cool climate regions. A trial block would be advisable before undertaking a large-scale planting.

A clone of 3309C free of leafroll virus 3 has recently been made available in Australia.

✘ *Not suitable for:*

It is not suited to dry and shallow conditions and not appropriate for heavy soils. It has some tolerance to lime (up to 11% active lime) and acid soils, but does not tolerate saline soils. It has a tendency to induce potassium deficiency in overcropped young vines on clay soils. Young vines grafted to 3309C can be very nutrient demanding (Cirami, 1999)

V. berlandieri × *V. riparia*

These rootstocks confer moderate to high vigour to the scion depending on the soil type. They can be moderately sensitive to drought conditions, but may hasten ripening and improve fruitset. They are best suited to cool climate viticulture, but with appropriate site selection and management they can produce quality outcomes in warmer regions. The varieties include:

- ➤ 5C Teleki
- ➤ 5BB Kober (5A Teleki) [3]
- ➤ SO4
- ➤ 420A

➤ 5C Teleki

5C Teleki accounts for less than 4% of total rootstock use in South Australia. It is widely used in Germany and is relatively popular in California and Victoria.

Characteristics:

- Moderate-high yields (Cirami, 1999)
- Moderate vigour – less than 5BB Kober and SO4 (Cirami, 1999)
- Moderately susceptible to drought conditions, similar water requirements to own roots
- More salinity-tolerant than own roots (Tee and Burrows, 2004)
- Moderate resistance to root-knot and root-lesion nematode but low resistance to citrus and dagger nematode (Whiting, 2004 and Nicol et al., 1999)
- Can suffer from magnesium deficiency (N. Dry, unpublished.)
- Moderate lime tolerance (Whiting, 2004)
- May advance maturity—it is the earliest maturing of the group (Galet, 1998)
- May improve fruitset (Cirami, 1999)

✔ *Suitable for:*

Moist free-draining, fine textured, calcareous soils with moderate depth and fertility, where water extraction during spring is slow and water availability over the ripening period is more consistent.

5C Teleki is best suited to cool climate viticulture as it imparts moderate vigour to the scion, may advance maturity and improve fruitset. It has performed particularly well when grafted to Cabernet Sauvignon in Coonawarra and Langhorne Creek, Shiraz in Wrattonbully and Chardonnay and Pinot Noir in the Adelaide Hills.

✘ *Not suitable for:*

5C Teleki should not be used at sites that do not have adequate supplies of good quality water. While 5C Teleki could be used in the Riverland (so long as there was adequate water) there are more appropriate rootstocks.

In deep sand and sandy loam soils it does have a tendency to produce excessive vigour in spring and can suffer from early senescence once the soil dries out in late

---

3. Treated as synonyms. See page 11.

summer (N. Dry, unpublished). Its use under these circumstances should therefore be avoided. 5C Teleki also does not perform well on acid soils that have not been ameliorated (Whiting, 2003).

➤ 5BB Kober/5A Teleki

5BB Kober and 5A Teleki have been grown in Australia for many years. The two varieties were recently found to be genetically identical, although there seem to be some clonal differences. The name 5BB Kober is used here to refer to both clones. 5BB Kober accounts for only 3% of rootstock plantings in South Australia but it is widely used in Germany.

Characteristics

- Moderate-high yields (Cirami, 1999)
- Moderate-high vigour (Cirami, 1999)
- Drought tolerance is site-dependent—generally more tolerant than 5C Teleki and own roots
- More salinity-tolerant than own roots (Tee and Burrows, 2004)
- High resistance to root-knot but low resistance to citrus, dagger and root-lesion nematode (Whiting, 2004 and Nicol et al., 1999)
- High lime tolerance—up to 20% active lime (Galet, 1998)

✔ *Suitable for:*
Moist, fine textured, calcareous soils of moderate depth and low to moderate fertility where water extraction during spring is slow and water availability over the ripening period is more consistent.

5BB Kober is best suited to low to moderate potential sites in cool climates and has performed particularly well when grafted to Cabernet Sauvignon in Coonawarra, with Chardonnay in Wrattonbully and with Riesling in Eden Valley.

✘ *Not suitable for:*
In deep sand and sandy loam soils 5BB Kober, like 5C Teleki, does have a tendency to produce excessive vigour in spring and can suffer from early senescence once the soil the dries out in late summer (N. Dry, unpublished). Excessive vigour can also be a problem in fertile soils.

There have been reports of incompatibility of certain *V. vinifera* varieties (e.g. Chardonnay) with 5BB Kober. Local experience indicates that any scion with a significant virus load grafts poorly with this rootstock, which points to the need to use certified material of known virus status.

➤ SO4

SO4 has been widely used in Germany and France because it is easy to graft and adapts well to a range of soil types. Its use in South Australia has been limited (less than 1% of total rootstock plantings).

Characteristics

- High-moderate yields (Cirami, 1999)
- Moderate-high vigour, depending on the site
- Performs poorly in drought and saline conditions (Southey, 1992)
- Moderate-high resistance to root-knot nematode, moderate resistance to root-lesion nematode and low resistance to citrus and dagger nematode (Whiting, 2004 and Nicol et al., 1999).
- Moderate-high lime tolerance—active lime of up to 17–18% (Galet, 1998)
- May advance maturity if vigour is controlled
- May improve fruitset if vigour is controlled
- Performs satisfactorily in acid soils

✔ *Suitable for:*

A wide range of soils but does best in sandy, well drained soils of low fertility. Based on overseas experience SO4 is best suited to moderate potential sites in cool climate regions.

✘ *Not suitable for:*

It does not tolerate drought conditions. It is moderately sensitive to salinity and so should not be used at sites where there are not adequate supplies of good quality water. Use of SO4 in deep, fertile soils is likely to lead to excessive vine vigour.

SO4 assimilates magnesium poorly and this is reported to contribute to bunch necrosis in Merlot and Cabernet Sauvignon under French conditions (Galet 1998). It should be noted that SO4 has been criticised for its slender trunk, which can cause breakage at the grafts with mechanical harvesting. In addition, it is reported that vigour decreases considerably after 15–20 years (Galet, 1998).

➤ 420A

Has not been used for any commercial vineyards in South Australia but is used for quality wine production in France and California.

Characteristics:

- Low-moderate vigour (Whiting, 2003)
- Low-moderate yield (Whiting, 2003)
- Susceptible to drought and saline conditions (Southey 1992)
- Moderate resistance to root-knot nematode but low resistance to citrus, root-lesion and dagger nematode (Whiting, 2004 and Nicol et al., 1999)
- High lime tolerance—up to 20% active lime (Galet, 1998)
- Shallow growing and well-branched root system
- Long vegetative cycle
- May increase fruitset (Candolfi-Vasconcelos, 1995)

✔ *Suitable for:*

Considering its low vigour and lime tolerance it may be a good rootstock for high quality wine grape production in the heavy-textured, calcareous soils of Coonawarra. There is little experience in Australia so a trial block would be advisable before undertaking a large-scale planting.

In a trial at Banksdale, Victoria with Merlot, 420A produced the best overall mineral balance, in terms of low potassium and adequate magnesium and calcium compared with SO4, 99 Richter, 101–14 and Schwarzmann (M. Walpole, pers. comm.).

✘ *Not suitable for:*

Avoid use in warm and hot regions as it performs poorly under dry conditions and has low salinity tolerance. It also performs poorly in waterlogged soils (Whiting 2003) and is susceptible to *phytophthora* (Southey, 1992).

V. berlandieri × *V. rupestris*

These rootstocks impart moderate-high vigour to the scion, are drought-tolerant and have moderate-high tolerance of nematodes and saline soils. They have a long vegetative cycle and they are best adapted to warm-hot regions and can produce both commercial quality at high yields or lower yielding premium quality wine grapes. The varieties include:

- ➤ 1103 Paulsen
- ➤ 140 Ruggeri
- ➤ 110 Richter
- ➤ 99 Richter

➤ **1103 Paulsen**

1103 Paulsen has been the most widely planted rootstock in South Australia in the last 10 years (1997–2006) and accounts for 17% of total rootstock plantings. 1103 Paulsen has also been widely used in North Africa and the dry, arid regions of Italy and France.

Characteristics

- Moderate-high yields
- Moderate-high vigour
- Deep growing, strongly developed root system (Cirami, 1999)
- Excellent drought tolerance and high water-use efficiency
- Excellent salinity tolerance (Tee and Burrows, 2004)
- Moderate-high resistance to root-knot nematode, moderate resistance to citrus and root-lesion nematode and low resistance to dagger nematode (Whiting, 2004 and Nicol et al., 1999)
- Moderate-high lime tolerance—up to 17–18% active lime (Galet, 1998)
- Long vegetative cycle which may delay maturity

✔ *Suitable for:*

1103 Paulsen is best suited to low and moderate depth soils and a wide range of soil types including sand, sandy loams and clays. It can tolerate a moderate degree of waterlogging (it is superior to 140 Ruggeri in this regard) and has performed well in the acid soils of Eden Valley.

Its drought and salinity tolerance make this rootstock particularly well suited to warm and hot regions. In the Riverland it has been used extensively with Chardonnay, Shiraz and Cabernet Sauvignon and consistently produced relatively high yields of good quality. In the Barossa Valley and McLaren Vale it has produced premium quality outcomes with Shiraz and Cabernet Sauvignon, in both sandy-loams and red-brown earths. In Eden Valley and Clare Valley it has also been used successfully with Riesling and tends to hold onto its leaves for longer compared with lower vigour rootstocks and own roots.

While it is better suited to warm-hot climates with appropriate selection and management it can produce quality outcomes in cooler regions (more so with white rather than red varieties).

The key to producing quality with this rootstock is to manage the vigour through appropriate site selection (low-moderate potential sites) and low inputs of water and nitrogen. Where vigour is not controlled quality can suffer.

Growers have reported that this rootstock can 'start off a bit slow' after establishment, but does 'catch-up' in time. This has implications for nitrogen application; growers should avoid the temptation of overdosing early as it may lead to vigour problems as the vines mature.

✘ *Not suitable for:*
Do not use this rootstock in deep, fertile soils as it will lead to the production of excessive vigour. 1103 Paulsen has a long vegetative cycle which may delay ripening in cool climates. There have been isolated incidences of poor performance in association with aggressive root-knot nematode populations in the Riverland.

➤ 140 Ruggeri

140 Ruggeri is the third-most-planted rootstock in South Australia accounting for 9% of total rootstock plantings. As with 1103 Paulsen it is widely planted in North Africa and the dry, arid regions of Italy and France.

Characteristics

- Moderate-high yields
- Moderate-high vigour—generally higher than 1103 Paulsen but less than Ramsey
- Excellent drought tolerance and high water-use efficiency
- Good salinity tolerance (Tee and Burrows, 2004)
- High lime tolerance—up to 20% active lime (higher than 1103 Paulsen)
- It has high resistance to root-knot nematode but low resistance to root-lesion and citrus nematode (Whiting, 2004 and Nicol et al., 1999)
- Long vegetative cycle, which may delay ripening

✔ *Suitable for:*
140 Ruggeri is a very hardy rootstock suitable for the most challenging drought conditions. It is also able to tolerate lime and is adapted to acid soils. It is similar to 1103 Paulsen in that it performs best in low to moderate depth soils across a range of soil types.

Its drought and salinity tolerance make this rootstock particularly well suited to warm and hot regions. In the Riverland it has been used extensively with Chardonnay, Shiraz and Cabernet Sauvignon and has consistently produced relatively high yields of good quality. In the Barossa Valley and McLaren Vale it has produced premium quality outcomes with Shiraz and Cabernet Sauvignon in both sandy-loams and red-brown earths.

Its high resistance to root-knot nematode would make 140 Ruggeri the preferred option to 1103 Paulsen in sandy soils.

While it is better suited to warm-hot climates, with appropriate selection and management it can also produce quality outcomes in cooler regions (more so with white rather than red varieties).

The key to producing quality with this rootstock is to manage the vigour through appropriate site selection (low-moderate potential sites) and low inputs of water and nitrogen, where vigour is not controlled quality can suffer.

✘ *Not suitable for:*

This rootstock is moderately susceptible to spring waterlogging particularly in its formative years. Do not use this rootstock in deep, fertile soils as this will lead to the production of excessive vigour. 140 Ruggeri has a long vegetative cycle which may delay ripening in cool climates.

➤ 110 Richter

110 Richter is extensively used in the dry, warm-hot regions of Portugal, Spain, Greece, North Africa and France (Cirami, 1999). Despite its suitability to local conditions it accounts for less than 2% of rootstock plantings in South Australia. Its low uptake may be explained by nursery reluctance to recommend this rootstock due to its unpredictable nursery propagation results (Galet, 1998).

Characteristics

- Moderate yield
- Moderate vigour
- Excellent drought tolerance and high water-use efficiency
- Tolerates salinity better than own roots (Tee and Burrows, 2004)
- Moderate-high lime tolerance—up to 17% active lime (Galet, 1998)
- It is moderately resistant to root-knot and citrus nematode, but has low resistance to root-lesion and dagger nematode (Whiting, 2004 and Nicol et al., 1999)
- Tolerates waterlogged soils better than most rootstocks (Southey, 1992)
- Long vegetative cycle, which may delay ripening.

✔ *Suitable for:*

A rootstock suitable for most soils from slightly acidic through to calcareous soils. It does well on badly drained, shallow clay soils that can suffer from both waterlogging and drought conditions (Whiting, 2003).

Trials indicate that this rootstock produces adequate yield in a variety of situations while vine balance and fruit quality are reported to be excellent. 110 Richter is well suited to premium wine production in warm regions. Because it imparts moderate vigour to the scion, compared with 140 Ruggeri and 1103 Paulsen, it is a good alternative on moderate-high potential sites. Growers in Langhorne Creek have reported that 110 Richter performs similarly to ungrafted Shiraz, Cabernet Sauvignon and Chardonnay in terms of vigour and yield but requires less water.

In the Riverland it has performed well on heavier textured soils with Shiraz. Because it has excellent drought tolerance and some salinity tolerance but imparts moderate vigour, it should be a good long term option for cooler regions such as Padthaway and Coonawarra.

As with 99 Richter it can be a slow starter in the first year as it expends its energy on root growth (Galet 1998).

✘ *Not suitable for:*

It may delay ripening or cause excessive vigour on high potential sites in cooler regions. It has struggled to produce adequate canopies on sandy, low fertility soils in the Riverland. It is not suited to scion varieties with irregular set (Cirami 1999).

➤ 99 Richter

99 Richter accounts for less than 1% of rootstock plantings in South Australia, but has been used extensively in Victoria and South Africa. In France the use of this rootstock has been abandoned in favour of other *V. berlandieri* × *V. rupestris* rootstocks (Galet, 1998).

Characteristics

- Moderate-high yield (Whiting, 2003)
- Moderate-high vigour
- Some drought tolerance—less than other *V. berlandieri* × *V. rupestris* rootstocks
- More salinity-tolerant than own roots (Tee and Burrows, 2004)
- 99 Richter has moderate-high resistance of root-knot nematode, but low resistance to root-lesion, citrus and dagger nematode (Whiting, 2004 and Nicol et al., 1999)
- Moderate-high lime tolerance—up to 17% active lime (Galet, 1998)

Suitable for:

It has good tolerance of lime and performs well in acid soils. As with other rootstocks within this family it has a degree of drought tolerance; however, it is the least effective of the family.

✘ *Not suitable for:*

In the first year after planting growth may be initially slow, particularly in cold soils. It is not suited to soils that are prone to spring waterlogging (Whiting and Orr, 1990). It can produce excessive vigour so should not be used in high potential sites (deep, moist soils).

V. champini

➤ Ramsey

Ramsey is the most widely planted rootstock in South Australia accounting for 36% of rootstock plantings. However in the past 10 years (1997–2006) it has been surpassed by 1103 Paulsen in terms of cutting sales.

Characteristics

- High yield
- High vigour
- Excellent drought tolerance and high water-use efficiency
- Good salinity tolerance (Tee and Burrows, 2004)
- High resistance to all nematode species except for dagger nematode (Whiting, 2004 and Nicol et al., 1999)
- Increased potassium levels are common in black varieties (Walker, 1998)
- Moderate lime tolerance (Whiting, 2003)
- Long vegetative cycle, which may delay ripening

✔ *Suitable for:*

Ramsey performs well in a range of soils so long as they do not encourage excessive vigour. Ramsey has excellent drought tolerance, good salinity tolerance and high nematode resistance. These three factors coupled with its high productivity make Ramsey particularly well suited to the production of consistently high yields of commercial quality wine grapes in the Riverland. The key to quality outcomes is to avoid high potential sites and ensure that inputs of nitrogen and water are kept significantly lower than what would be normal for ungrafted vines.

While Ramsey is best suited to hot climates, with appropriate selection and management (low potential site and low inputs of water and nitrogen), it has produced premium quality outcomes in the Barossa Valley and McLaren Vale.

It is important to highlight that Ramsey has consistently used significantly less water than own roots and other rootstocks in trials and is therefore a good long term option at sites where water availability is not guaranteed.

✗ *Not suitable for:*

Ramsey can produce excessive vigour, delay fruit maturity and has high potassium uptake which can have a negative effect on pH and wine colour. Based on this Ramsey may not be suited to premium red wine production, although as indicated previously premium outcomes have been achieved with this rootstock.

Ramsey is not suited to soils prone to spring waterlogging.

CSIRO (Merbein) rootstocks

CSIRO has released three new low-to-medium vigour rootstocks (Merbein 5489, Merbein 5512 and Merbein 6262) that are protected under Plant Breeders Rights (PBR) legislation. The rootstocks are derived from the American species *Vitis berlandieri* and *V. cinerea* and were selected from 55 advanced selections from the CSIRO breeding program.

They were selected to address problems associated with the adoption by the wine grape industry of high-vigour rootstocks. These problems were identified as negative impacts on berry composition associated with high potassium uptake, high grape juice pH and unfavourable organic acid composition, as well as poor colour of red wine.

Using Shiraz as the scion variety, the Merbein rootstocks were compared to commonly used rootstocks (Ramsey, 1103 Paulsen and 140 Ruggeri) in a replicated trial (warm climate, sandy loam soil, overhead irrigation). It was found that the Merbein rootstocks accumulated significantly less potassium in berries leading to lower juice and wine pH, and thus there was less need for acid adjustment in winemaking when using the Merbein rootstocks. The Merbein rootstocks also had a positive effect on wine spectral properties.

These new low-to-medium vigour rootstocks are now available for industry to evaluate in other regions with a wider range of scion varieties. At the time of writing, the performance of these rootstocks had not been validated in commercial vineyards.

➤ Merbein 5489

Summarised performance with Shiraz under irrigation in a warm climate

- Relatively high yields
- Medium vigour
- Low juice potassium and pH reduces acid addition for pH adjustment in winemaking
- Enhances wine colour and phenolics

Merbein 5489 is a medium vigour rootstock. Pruning-wood weights from Shiraz vines grafted to Merbein 5489 were about half of those grafted to Ramsey and 1103 Paulsen in a Sunraysia-based trial. In the same trial, and despite having medium vigour, the Shiraz vines on Merbein 5489 produced similar yields per vine to those grafted to 1103 Paulsen rootstock.

In screening trials, Merbein 5489 was tolerant of the G1, G4 and G30 strains of phylloxera.

Shiraz vines grafted to Merbein 5489 established in a re-plant soil infested with nematodes showed no ill effects due to nematode infection over a period of 18 years. Vines of Merbein 5489 were resistant to one biotype of *Meloidogyne javanica* in a pot-trial conducted under glasshouse conditions. They were also tolerant of *Meloidogyne incognita* and a second biotype of *M. javanica* in that whereas galls and egg masses were spread throughout Sultana vine roots, they were either absent or observed at a low frequency in root systems of Merbein 5489.

Based on levels accumulated in leaf petioles at harvest under irrigation with River Murray water in Sunraysia, Merbein 5489 had an ability equivalent to 1103 Paulsen for sodium and chloride exclusion.

Shiraz vines grafted to Merbein 5489 had a higher crop water-use index than those grafted to 1103 Paulsen in the Sunraysia region. High crop water-use index (crop produced per unit of water used) is an indicator of water-use efficiency.

When grafted to Merbein 5489, Shiraz vines accumulated significantly less potassium in berries leading to a lower juice and wine pH and thus less need for acid adjustments in winemaking.

Wine made using fruit from Shiraz vines grafted to Merbein 5489 in the warm Sunraysia region had higher colour density, lower colour hue (brighter wine), higher total phenolics and higher ionized anthocyanins (the coloured form of anthocyanins) than equivalent wines made using fruit from Shiraz grafted to 1103 Paulsen and Ramsey rootstocks.

➤ Merbein 5512

Summarised performance with Shiraz under irrigation in a warm climate

- Moderate yields
- Low vigour
- Low juice potassium and pH reduces acid addition for pH adjustment in winemaking
- Enhances wine colour and phenolics.

Merbein 5512 is a low vigour rootstock. Pruning-wood weights from Shiraz vines grafted to Merbein 5512 were about one-third of those grafted to Ramsey and 1103 Paulsen in a Sunraysia-based trial. In the same trial, the Shiraz vines on Merbein 5512, with their reduced vigour, produced yields per vine which were 25 and 47% lower than those grafted to 1103 Paulsen and Ramsey rootstocks, respectively.

In screening trials, Merbein 5512 was tolerant of the G4 and G30 strains of phylloxera.

Shiraz vines grafted to Merbein 5512 established in a re-plant soil infested with nematodes showed no ill effects due to nematode infection over a period of 18 years. In comparison with own-rooted Sultana vines, Merbein 5512 was tolerant of *Meloidogyne javanica* and *Meloidogyne incognita* in a pot trial conducted under glasshouse conditions. Whereas galls and egg masses were spread throughout Sultana roots, they were either absent or observed at a very low frequency in root systems of Merbein 5512.

Based on levels accumulated in leaf petioles at harvest under irrigation with River Murray water in Sunraysia, Merbein 5512 had an ability equivalent to Ramsey and 1103 Paulsen for chloride and sodium exclusion, respectively.

When grafted to Merbein 5512, Shiraz vines accumulated significantly less potassium in berries leading to lower juice and wine pH and thus less need for acid adjustments in winemaking.

Wine made using fruit from Shiraz vines grafted to Merbein 5512 in the warm Sunraysia region had higher ionized anthocyanins (the coloured form of anthocyanins) and lower colour hue (brighter wine) than equivalent wines made using fruit from Shiraz grafted to 1103 Paulsen and Ramsey rootstocks. Wine from Shiraz on Merbein 5512 also had higher colour density and higher total phenolics than that from Shiraz grafted to Ramsey.

➤ **Merbein 6262**

Summarised performance with Shiraz under irrigation in a warm climate

- Moderate yields
- Low vigour
- Low juice potassium and pH reduces acid addition for pH adjustment in winemaking
- Enhances wine colour and phenolics

Merbein 6262 is a low vigour rootstock. Pruning-wood weights from Shiraz vines grafted to Merbein 6262 were about one-third of those grafted to Ramsey and 1103 Paulsen in a Sunraysia-based trial. In the same trial, the Shiraz vines on Merbein 6262, with their reduced vigour, produced yields per vine which were 16 and 41% lower than those grafted to 1103 Paulsen and Ramsey rootstocks, respectively.

In screening trials, Merbein 6262 showed moderate tolerance towards the G4 and G30 strains of phylloxera. Moderate tolerance indicates a risk of phylloxera developing on the roots but at relatively low levels compared to *Vitis vinifera.*

Shiraz vines grafted to Merbein 6262 established in a re-plant soil infested with nematodes showed no ill effects due to nematode infection over a period of 18 years. Vines of Merbein 6262 were resistant to two biotypes of *Meloidogyne javanica* in a pot trial conducted under glasshouse conditions. They were also tolerant of *Meloidogyne incognita* in that whereas galls and egg masses were spread throughout Sultana vine roots, they were either absent or observed at a low frequency in root systems of Merbein 6262.

Based on levels accumulated in leaf petioles at harvest under irrigation with River Murray water in Sunraysia, Merbein 6262 had an ability equivalent to 1103 Paulsen and Ramsey for sodium exclusion. However, mean concentrations on a dry weight basis for chloride accumulated in petioles of Shiraz vines were 0.95% grafted to Merbein 6262 compared with 0.34% grafted to Ramsey.

When grafted to Merbein 6262, Shiraz vines accumulated significantly less potassium in berries leading to lower juice and wine pH and thus less need for acid adjustments in winemaking.

Wine made using fruit from Shiraz vines grafted to Merbein 6262 in the warm Sunraysia region had higher colour density, lower colour hue (brighter wine), higher total phenolics and higher ionized anthocyanins (the coloured form of anthocyanins) than equivalent wines made using fruit from Shiraz grafted to 1103 Paulsen and Ramsey rootstocks.

Specialty rootstocks

> ➤ **Börner** – a rootstock with complete resistance to phylloxera

In the early 20th century Carl Börner discovered different biotypes of phylloxera. He also observed that no galls developed on the leaves or roots of *V. cinerea* (Arnold) and that it was possible to transmit this reaction by crossbreeding.

After many years of crossbreeding *V. cinerea* with *V. riparia* a new variety called Börner was developed. The new variety was then infested with phylloxera strains from Germany, Italy, France and the USA and it was found that no galls developed on the leaves or roots (Becker, 1988).

Börner rootstock is vigorous, winter hardy and produces yields similar to SO4 (Becker, 1988). Börner shows good adaptation to a wide range of soils, except for extremely calcareous soils (Becker and Wheeler, 2000) and when grafted to Riesling it performed equally or better than other rootstocks particularly under dry conditions. (Schmid et al., 1998). It has also been found to suffer from iron chlorosis under wet conditions and it has moderate-poor rooting ability. (Schmid et al., 1998)

Börner has very high phylloxera resistance and the nature of its resistance to phylloxera (hypersensitive reaction) would suggest that it would have good resistance to all species of nematodes.

When grafted to Riesling and compared with other rootstocks over a number of seasons, no negative effects in terms of quality were found (Becker, 1988).

In 2006 PGIBSA, in conjunction with the local vine improvement groups, established a trial in the Clare Valley comparing the performance of Riesling grafted to Börner, 110 Richter, SO4 and own roots and a trial in the Adelaide Hills comparing the performance of Chardonnay grafted to Börner, 5C Teleki, 110 Richter and own roots. Preliminary information on the suitability of Börner for Australian conditions will be available by 2010.

> ➤ **Gravesac** – a rootstock bred to tolerate acid soils

Gravesac was bred specifically to tolerate low pH (high acid) soils and is a cross of 161–49C and 3309C (*V. berlandieri* × *V. riparia* × *V. rupestris*) (Galet, 1998). There is very little information on this rootstock because it has not been widely planted in any viticultural regions.

Galet (1998) describes Gravesac as having high vigour, whereas Delas (1992) describes Gravesac as having moderate vigour. Gravesac advances maturity (Delas, 1992) and has high resistance to phylloxera (Galet, 1998).

Site Factors

There is no single rootstock that can meet the different challenges of every site. Each rootstock has its own particular set of characteristics, strengths and weaknesses. The selection of the most appropriate rootstock for any given situation requires a thorough understanding of the particular site in which the vines are to be planted. The following chapter discusses the site factors which influence rootstock performance and therefore rootstock selection.

Soil Properties

It is the soil properties (e.g. texture, exploitable depth, waterlogging potential and soil chemistry) that are relevant for rootstock selection rather than the soil type itself. These properties are determined by soil analysis and characterisation prior to planting.

Physical Properties

Soil texture and rootstock root distribution

Soil properties (physical and chemical) and cultural practices (soil management and irrigation) have a large influence on the root distribution of a grapevine (Smart et al., 2006); however, variation in root distribution also has a genetic component, i.e. is influenced by rootstock genotype (Swanpoel and Southey, 1989). The root distribution of a rootstock is a function of its parentage (Figure 1):

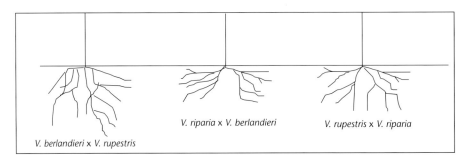

Figure 1. Hypothetical rootstock root distribution patterns adapted from Guillon (1905) and based on the emergence angles of American *Vitis* species.

- *V. berlandieri* × *V. rupestris* hybrids have a root system that is dense and penetrates deep into the soil
- *V. riparia* × *V. berlandieri* hybrids have a relatively shallow, lateral spreading root system
- *V. riparia* × *V. rupestris* hybrids have a relatively shallow, lower density root system

The wetting pattern beneath a dripper is influenced by the soil texture (Figure 2). The hypothetical interactions between wetting pattern and root distribution and the extent to which the roots of each rootstock genotype are found within each wetting pattern may help to explain rootstock performance on different soil types.

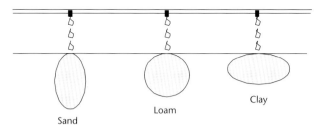

Figure 2. The influence of soil texture on the wetting pattern beneath a dripper.

The following section discusses these hypothetical interactions and the implications for rootstock selection and management. It must be noted that; the discussions are based on hypothetical assumptions that need to be confirmed by further research, the diagrams are general in nature and are for demonstration purposes only and within each rootstock hybrid group there are differences in root distribution.

V. riparia × *V. rupestris*

Rootstocks from the cross *V. riparia* × *V. rupestris* (101–14, Schwarzmann and 3309C) (Figure 3) have a more plunging root system compared with *V. riparia* × *V. berlandieri*. These are therefore more suited, and would achieve the maximum potential for water-

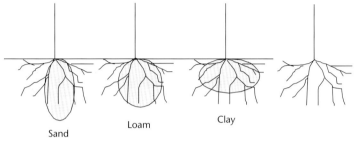

Figure 3. The hypothetical interactions between the root distribution pattern of *V. riparia* × *V. rupestris* rootstocks and the wetting patterns of sand, loam and clay soils.

use efficiency when planted on loam and clay-loam soils. These rootstocks can struggle in the heat of summer on sandy soils which is not surprising when considering the low proportion of roots in the wetting pattern of the sandy soil.

V. riparia × V. berlandieri

Rootstocks from the cross *V. riparia × V. berlandieri* (5C Teleki, 5BB Kober and SO4) are best suited to clay and to a lesser extent loam soils (Figure 4). This interaction may help to explain why these rootstocks struggle on sands in the heat of summer, i.e. water is not reaching the majority of the roots. One would also expect lower water-use efficiency on the sand and sandy-loam soils, with maximum potential for water-use efficiency achieved on clay soils.

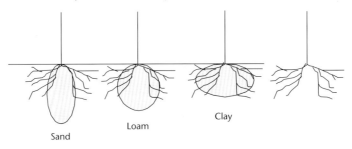

Figure 4. The hypothetical interactions between the root distribution pattern of V. *riparia* × V. *berlandieri* rootstocks and the wetting patterns of sand, loam and clay soils.

V. berlandieri × V. rupestris

Rootstocks from the cross *V. berlandieri × V. rupestris* (140 Ruggeri, 1103 Paulsen, 110 Richter and 99 Richter) with their plunging, dense root systems have a good proportion of roots in each of the three wetting patterns and so are suited to all soil types (Figure 5).

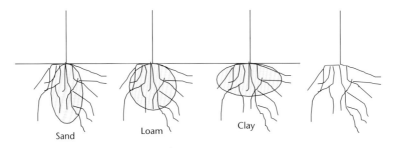

Figure 5. The interaction between the rooting pattern of V. *berlandieri* × V. *rupestris* rootstocks and the wetting pattern of sand, loam and clay soils.

Soil Depth

The depth to the impenetrable or impermeable layer is an important soil property as it determines the volume of soil that the vine roots can explore for moisture and nutrients. In general, the greater the potential rooting depth, the greater the potential for vigour. When planting on shallow soils (<40cm) a grower should consider using a more vigorous rootstock; and when planting on a deep soil (>80cm) a lower vigour rootstock should be considered.

Waterlogging Potential

Waterlogging occurs at sites where the movement of water into the soil surface or through the soil profile is impeded, or where there is a perched or rising water table. Waterlogged soils are best identified by observing the soil after heavy rain in late winter and early spring (Maschmedt et al., 2004).

All vines, whether they are grafted or not, will perform poorly in waterlogged soils, although some varieties and rootstocks perform better than others. Ideally growers should avoid planting into these areas or amend the soil prior to planting. Depending on the type of impediment this can be achieved by mounding soil, constructing agricultural drains, or deep ripping with gypsum (Maschmedt, 2004). Growers should note that, in general, ungrafted vines perform better than rootstocks in waterlogged soils. Those rootstocks that are known to be most sensitive to waterlogged conditions are:

- 99 Richter
- Ramsey
- 140 Ruggeri
- 420A

Those that are classed as the least sensitive are:

- Schwarzmann
- SO4
- 110 Richter
- 1103 Paulsen

- 101–14 (*susceptible in early years, but more tolerant as the vine develops*)

Reference for rootstock classifications:
www.sardi.sa.gov.au/pages/hort/viticulture/rootstock_characteristics.htm

Chemical Properties

Soil Acidity

Soils below pH 5.5 can cause nutrient deficiency (N, P, K, S, Ca and Mg) or conversely nutrient toxicity (Al, Cu and Mn) in grapevines. There are very few areas in South Australia where growth is limited by acidic soils and as a result there has been little local research done on rootstock performance under such conditions. The only region in South Australia with widespread soil acidity issues is the Adelaide Hills. Where soil acidity is an issue growers should ameliorate with lime to correct any pH imbalance before establishment and avoid using acidifying ammonium-based fertilisers. If the aforementioned management strategies are unable to permanently correct soil pH then growers should use the following rootstocks:

- 140 Ruggeri
- 1103 Paulsen
- 99 Richter
- 110 Richter

Avoid using the following rootstocks in acid soils that have not been ameliorated:

- Schwarzmann
- 101–14
- SO4
- Ramsey
- 5C Teleki
- 5BB Kober
- 3309C   (Whiting, 2003)

Soil Lime Content

Many vineyards in South Australia are planted on calcareous (high lime content) soils. These soils have the potential to contribute to lime-induced chlorosis, particularly in cold, wet, springs. Growers need to be aware that vines grafted to rootstocks are more susceptible to lime-induced chlorosis than ungrafted vines. Lime-induced chlorosis, however, does not appear to be a major economic issue in grafted South Australian vineyards. In a study of 36 different rootstock blocks across six regions none experienced economic losses due to lime induced chlorosis, despite a large proportion being situated on calcareous soils (N. Dry, unpublished). There is, however, anecdotal evidence of problems in some parts of the Riverland and Wrattonbully. Those rootstocks that can tolerate the highest levels of active lime are descended from *Vitis berlandieri*; specifically*:

- 140 Ruggeri
- 5BB Kober
- 420A
- 110 Richter
- 1103 Paulsen
- 99 Richter
- SO4
- Ramsey
- 5C Teleki

This is in order from most to least lime tolerant according to Galet's (1998) rankings. Ramsey and 5C Teleki were not included in Galet's ranking but have been classified as moderately tolerant in Whiting (2004).

Nutrient Elements

Nitrogen and potassium are the two most important nutrient elements that growers need to consider when selecting and managing rootstocks.

Nitrogen

In general most rootstocks take up and assimilate nitrogen more efficiently than own-rooted vines and in the past the excessive vigour imparted by grafted vines was often a result of growers fertilising grafted vines at the same rate as own-rooted vines. Research has also shown that some rootstocks are more responsive to changes in nitrogen supply than others (Keller et al., 2001a). For further information on nitrogen/rootstock interactions and the implications for management go to page 75.

Potassium

Rootstocks influence potassium uptake and juice potassium levels. May (1994) states that; high levels of potassium uptake have been shown to increase juice pH above desired levels leading to wine instability and poor colour in red wines. Most Australian viticultural soils are abundant in potassium and so rootstock influence on juice potassium levels may be an important selection criteria. Rootstocks with *Vitis champini* parentage (Ramsey, Freedom, Harmony Dog Ridge, K51–32 and K51–40) have comparatively high juice potassium levels and so are not generally used in Australian viticulture (except for Ramsey). According to Whiting (2003) the following rootstocks tend to have moderate juice potassium levels:

- Schwarzmann
- 140 Ruggeri
- 99 Richter
- 101–14

Rootstocks that have lower juice potassium levels are:

- 420A
- 110 Richter
- 5C Teleki
- 5BB Kober
- 1103 Paulsen
- SO4

Note: the CSIRO rootstocks released in 2006 have all been bred for low potassium uptake – see page 19.

Soil Fertility

The vigour potential of a soil increases with soil fertility. It is possible to predict soil fertility in two ways: either by examining soil texture or by using cation exchange capacity (CAC) data from soil tests.

Soil Texture

Generally, soil fertility increases with increasing soil clay content as clay soils have a greater capacity to store nutrients compared with sand (Gladstones, 1992).

Cation Exchange Capacity (CEC)

Cation exchange capacity (CEC) is a measure of the inherent fertility of a soil i.e. the soils capacity to store and release major nutrient elements (Maschmedt et al., 2004). The higher the CEC of each soil layer, the higher the inherent fertility of the soil (Table 1).

Table 1. Inherent soil fertility classified by the sum of cations in each soil layer. Information based on PIRSA Land Information (2000).

Sum of cations in each layer	Inherent fertility
< 5	Low
5–15	Moderate
15–25	High
> 25	Very high

Table 2 highlights where to find the CEC data on a soil analysis test result and Tables 3, 4 and 5 are examples from actual vineyard soil analyses.

Table 2. Typical layout of data in soil test. The column relevant to the estimation of a soil's inherent fertility (sum of cations) is circled.

Depth cm	pH H₂O	pH CaCl₂	CO₃ %	EC 1:5 dS/m	EC_e dS/m	Org C %	Avail P mg/kg	Avail K mg/kg	Cl mg/kg	SO₄-S mg/kg	Boron mg/kg	React Fe mg/kg
0–15	6.4	5.8	0	0.180	1.95	0.90	57	229	157	36.8	0.5	477
15–40	6.7	6.1	0	0.038	0.60	0.21	20	160	24	8.1	0.4	434
40–85	6.9	6.2	0	0.043	0.78	0.14	2	180	32	7.4	0.3	320
85–110	8.1	7.1	0	0.274	2.35	0.40	2	419	151	72.5	3.0	831
110–150	8.7	7.6	0.3	0.339	2.31	0.28	2	460	191	59.3	3.6	862

Depth	Trace Elements mg/kg (EDTA)				Sum Cations cmol	Exchangeable Cations cmol (+)/kg				Est. ESP
cm	Cu	Fe	Mn	Zn	(+)/kg	Ca	Mg	Na	K	
0–15	4.06	74	56.5	1.86	6.1	4.59	0.55	0.48	0.46	7.9
15–40	1.04	35	51.9	0.20	2.7	1.88	0.22	0.20	0.36	na
40–85	1.02	18	25.3	0.08	2.3	1,42	0,26	0.28	0.36	na
85–110	4.00	67	114	0.23	17.2	5.83	7.46	2.87	1.00	16.7
110–150	4.86	79	147	0.33	20.5	8.19	6.54	4.71	1.07	23.0

Growers should avoid using high vigour rootstocks on soils with high inherent fertility (high clay content or high CEC in each layer). It is possible, but not advisable, to use low vigour rootstocks on soils with low inherent vigour as they may require the addition of fertiliser inputs to ensure optimum growth. For more information on using the cation exchange capacity to determine inherent soil fertility, refer to Section 3.2.2 in *Viticulture Volume 1: Resources* (Dry and Coombe, 2004).

Tables 3, 4 and 5: Examples of low, moderate and high fertility soils based on the CEC data from actual vineyard soil analysis.

Table 3. Low Fertility

Depth (cm)	Sum of cations cmol (+)/kg
0–15	3.4
15–35	1.3
55–70	1.3
70–100	10.5
885–115	7.5
100–160	6.5

Table 4. Moderate Fertility

Depth (cm)	Sum of cations cmol (+)/kg
0–15	9.6
15–32	7.9
32–60	24.9
60–85	26.2
85–115	22.4
115–160	25.8

Table 5. High Fertility

Depth (cm)	Sum of cations cmol (+)/kg
0–18	32.6
18–40	30.4
40–70	23.8
70–112	15.2
112–160	13.4

Drought Tolerance and Water Availability

Characterising the water-use efficiency or drought tolerance of the commonly used rootstocks in Australia can be a difficult exercise as there are many physiological mechanisms involved. These mechanisms interact with one another and are influenced by the scion variety, the soil, the environment and cultural practices.

Because of these interactions, the results and conclusions obtained from different studies on the water-use efficiency or drought tolerance of different rootstocks can be contradictory, making it difficult to definitively predict a rootstock's drought tolerance at a given site. A good rule of thumb is to remember that drought tolerance is related to vine vigour and generally the most vigorous vines (regardless of the rootstock) have the most extensive root systems and are usually the most drought-tolerant (Soar, 2004). The classification of rootstocks on the page opposite is based on the consistency of their performances in a number of studies from both Australia and overseas.[4]

---

4. Carbonneau, 1985, Cirami et al., 1994, Ezzahouani, and Williams, 1995, Gibberd et al., 2001, McArthy et al., 1997, Pech et al. 2001, Soar et al. 2006, Southey, 1992, Walker, 2004, Virgona et al., 2003

Group 1. (Highly Tolerant)
- Ramsey
- 140 Ruggeri

Group 2. (Tolerant)
- 1103 Paulsen
- 110 Richter
- 99 Richter

Group 3. (Moderately Susceptible)
- 5BB Kober
- 5C Teleki
- SO4

Group 4. (Susceptible)
- 101–14
- Schwarzmann
- 3309C

Growers should use drought-tolerant rootstocks if they:
- have soils with low readily available water values (< 50 mm)
- currently have or expect to have seasons where water availability is restricted
- have areas of the vineyard which suffer from loss of yield and quality as a result of not being able to get around their irrigation shifts in the peak water-use period

Because different rootstocks have different levels of drought tolerance they need to be managed differently. For further information on managing rootstocks according to their relative drought tolerance go to page 72.

Salinity

The salinity of applied irrigation water and salinity associated with soils or rising water tables can affect productivity in grapevines and is detrimental to wine quality. Grafting to salinity-tolerant rootstocks will reduce these effects as they accumulate salinity at lower levels than own-rooted vines (*Vitis vinifera*).

Soil salinity levels above 1.8 dS/m in the root-zone will restrict root growth and performance of own-rooted vines. Therefore where levels exceed this value salinity-

Table 6. A guide to the soil salinity tolerance of a range of rootstocks.
(from Tee and Burrows, 2004)

Classification of salt tolerance	Grapevine	Approximate threshold soil saturation paste salinity (ds/m)
Sensitive	own roots, 3309C, 1202C, K51–40	1.8
Moderately sensitive	5BB Kober, 5C, Teleki, 110 Richter, 99 Richter, K51–32	2.5
Moderately tolerant	140 Ruggeri, Schwarzmann, 101–14, Ramsey	3.3
Tolerant	1103 Paulsen	5.6

tolerant rootstocks should be used to ensure maximum yield and quality. Table 6 is a guide to the salinity tolerance of a range of commonly used rootstocks. The values refer to the approximate soil saturation paste salinity level (dS/m) above which yield reductions will occur. The reduction in yield may be as much as 15% for every one unit dS/m increase above the threshold (Zhang et al., 2002).

Irrigating with saline water will increase root-zone salinity. The salinity levels in the root-zone will depend on the salinity of the applied water, rainfall, evaporation, irrigation efficiency (leaching fraction or the % of water that moves past the bottom of the root-zone) and leaching efficiency. When choosing a rootstock for saline conditions growers need to consider current and future water and soil salinity levels.

Nematodes

Nematodes are present in all viticultural regions of South Australia (Nicol et al., 1999). Loss of production due to nematode infestation in Australian viticulture has been estimated at approximately 7% (Stirling et al., 1992) and there can be up to 60% loss in heavily infested vineyards (Nicol and van Heeswijck, 1997).

Nematodes are most commonly found in sandy soils or soils previously planted to grapevines or other horticultural crops. There are a number of different genera (types) of nematodes found in South Australian vineyards.

Root-knot nematode (Meloidogyne *spp.)*

Root-knot nematode is the most widely distributed genus with almost all vineyards on sandy soils infested. There are four species of root-knot nematode commonly found in South Australian vineyards: *M. javanica, M. hapla, M. incognita* and *M. arenaria. M. javanica* is the most common species found in the warmer vineyard regions and *M. hapla* is the most common species in cool climate regions (Nicol et al., 1999).

Citrus nematode (Tylenchulus semipenetrans)

Citrus nematode, as the name implies, is commonly found in vineyards planted on old citrus ground or in areas with widespread citrus plantings. It is more prevalent in fine (clay) rather than coarse (sand) textured soils.

Root-lesion nematode (Pratylenchus *spp.)*

Root-lesion nematode is less common in S.A. vineyards than root-knot nematode and citrus nematode but has been found in all major viticultural regions (Nicol et al., 1999). *P. zeae* is the most commonly occurring species of root-lesion nematode.

Dagger nematode (Xiphinema *spp.)*

Dagger nematode (*Xiphinema index*) is significant because it is a vector for grape fanleaf virus; however, it is only found in small areas in north-east Victoria. *X. americanum* is found in South Australian vineyards; however, there is little information on the damage levels and relative levels of rootstock tolerance to this species.

> **In nematode-infested soils rootstocks are the most effective management option.**

Different rootstocks have different levels of tolerance to different nematodes and tolerance to one genus or species does not imply tolerance to all. Nematode tests pre-planting to determine the nematode genus and species and population density are therefore highly recommended.

When assessing the results from the nematode test it is important to remember that nematodes multiply rapidly. In ideal conditions one root-knot nematode juvenile can produce 125 million progeny in 3–4 months (Nicol et al., 1999). So, even if the results show that the current population levels are low, numbers will increase once vines are planted, causing an economic impact later in the vineyard's development.

Rootstock Nematode Resistance

Table 7 can be used to determine which rootstock will be most effective against the different nematode populations.

Table 7. Classification of rootstock nematode resistance. The classifications for root-knot, lesion and dagger nematodes are based on Whiting (2004) and the classifications for citrus nematode are based on Nicol et al. (1999).

Nematode Resistance				
Rootstock	Root-knot *Meloidogyne* spp.	Citrus (*Tylenchulus semipenetrans*)	Root lesion (*Pratylenchus* spp.)	Dagger (*Xiphinema* spp.)
Ramsey	High	High	High	Low
Schwarzmann	High	High	Low	High
101–14	High	Low	Low	Low-Moderate
5C Teleki	Moderate	Low	Moderate	Low
5BB Kober	High	Low	Low	Low
SO4	Moderate-High	Low	Moderate	Low
1103 Paulsen	Moderate-High	Moderate	Moderate	Low
110 Richter	Moderate	Moderate	Low	Low
99 Richter	Moderate-High	Low	Low	Low
140 Ruggeri	High	—	Low	Low
3309C	Low	Low	Moderate	Moderate
420A	Moderate	Low	Low	Low

This table is a guide only; growers need to be aware that rootstock nematode resistance can vary depending on the nematode species, past cropping history and cultural management. More extensive information on rootstocks and nematodes can be found in Nicol et al. (1999)

Climatic Conditions

Influence on rate of ripening

Rootstocks may advance or delay ripening compared with own-rooted vines. There a number of reasons why a grower may choose to use rootstocks which influence the rate or onset of ripening. These decisions are based on the climatic conditions of the region, the conditions in particular sections of a vineyard site or the availability of labour at harvest.

- A grower may wish to advance the ripening of a whole vineyard in regions prone to inclement weather during the ripening period in order to lower the risk of late season disease development and/or to ensure ripeness before the end of the ripening period.

- A grower may wish to advance or delay the maturity of a particular section of vineyard so that it better matches the ripening of the major part of the vineyard.

- Alternatively growers may wish to delay or advance the maturity of certain blocks to spread the harvest load.

A three-year study in Sunraysia (Krstic and Hannah, 2004) found that 101–14 consistently advanced ripening of Chardonnay, Cabernet Sauvignon and Shiraz by at least a week compared with all other rootstocks in the study (5C Teleki, 5BB Kober, 1103 Paulsen and Ramsey). 5C Teleki has consistently advanced ripening of Cabernet Sauvignon in Coonawarra by a week compared with ungrafted Cabernet Sauvignon.

The following rootstocks may advance fruit maturity:

- 101–14
- Schwarzmann
- 3309C
- 420A
- 5C Teleki

And the following rootstocks may delay fruit maturity:

- Ramsey
- 140 Ruggeri
- 1103 Paulsen
- 110 Richter
- 99 Richter

Important note: the effect of rootstock on ripening and maturity is dependent on vine yield and site conditions. While Ramsey, 140 Ruggeri, 1103 Paulsen, 110 Richter and 99 Richter generally delay maturity, in dry conditions their drought-tolerant qualities may have a positive impact on canopy health and rate of ripening. In the same dry conditions vines grafted to drought susceptible rootstocks (101–14, Schwarzmann) may suffer from water stress leading to poor leaf condition and delayed maturity. This is a good example of thinking of the 'bigger picture' when it comes to rootstock selection.

Fruitset

Poor fruitset results from cool, wet and windy conditions during the pre-flowering and flowering period (May, 2004). Some varieties are also susceptible to poor fruitset in particular Merlot and to a lesser extent Chardonnay and Cabernet Sauvignon. Appropriate rootstock selection at sites that experience these conditions or with susceptible scion varieties will:

- Reduce the incidence of poor fruitset leading to more consistent yields
- Reduce the proportion of 'shot' berries which may increase quality (May, 2004)

Table 8. Rootstocks and their influence on fruitset. Australian findings in **bold**.

Classification	Rootstock	References
Improves fruitset and/or fertility	5C Teleki	**Cirami, 1999**, Candolfi-Vasconcelos, 1995
	Schwarzmann	**Cirami, 1999, Whiting, 2003,** Candolfi-Vasconcelos, 1995
	101-14	**Cirami, 1999**, Candolfi-Vasconcelos, 1995
	SO4	Delas et al., 1991
	3309C	Delas et al., 1991
	420A	Candolfi-Vasconcelos, 1995
	Riparia Gloire	Delas et al., 1991
Reduces fruitset/ not recommended for poor set varieties	1103 Paulsen	**Whiting, 2004**
	110 Richter	Candolfi-Vasconcelos, 1995

The most commonly used rootstock for this purpose in South Australia is 5C Teleki; however, growers have reported improvements with other rootstocks including 101–14 and Schwarzmann. Table 8 summarises findings from Australian and overseas research.

Vineyard Variability

All vineyards have variability in vine performance due to differences in soil type and/or topography throughout the block (Bramley, 2005). Vineyard variability results in:

- Inefficiencies in management inputs
- Inaccurate/less reliable yield forecasting
- Reductions in the consistency of grape quality   (Bramley and Hamilton, 2005)

Vineyard managers in Australia tend to manage variability *post-establishment* using the principles of precision viticulture. Rootstocks, however, offer vineyard managers the opportunity to manage vineyard variability *at establishment* by matching the potential vigour of a rootstock to the potential vigour of the soil in an attempt to increase vineyard uniformity.

Growers can do this in two ways:

- Match rootstock to soil type and set-up each rootstock as a separate management unit with the aim of uniform management of the vineyard.

- Match rootstock to soil type within a block or down a row. Figure 6 is an example from the Adelaide Hills where a grower has attempted to match rootstock vigour with the potential vigour of the soil based on the notion that site potential increases down a slope.

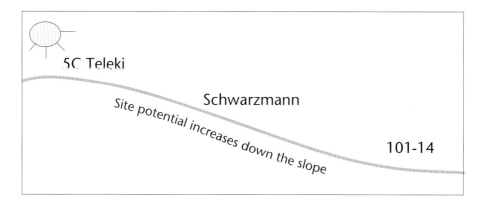

Figure 6. Schematic diagram of a vineyard layout where the grower has attempted to match rootstock vigour to site potential.

Table 9. Data from the 2007 vintage comparing the performance of rootstocks that have been matched to site potential. Values with same letter are not significantly different < 0.05.

Rootstock	5C Teleki	Schwarzmann	101–14
Inherent rootstock vigour	Moderate	Low-moderate	Low
Shoot number	15.07a	13.30b	14.07ab
Pruning weight (kg/vine)	0.45a	0.33b	0.42ab
Yield (kg/vine)	2.18a	1.23b	1.36b

The data in Table 9 demonstrate the ability of rootstocks to increase uniformity. It shows that despite differences in site potential from the top to the bottom of the hill, there was no significant difference in shoot number or pruning weight between the top section (5C Teleki) and the bottom section (101–14) and no significant difference in shoot number, yield and pruning weights between the middle section (Schwarzmann) and the bottom section (101–14). While this has not worked perfectly, i.e. no significant difference across the whole block, one needs to consider that if the grower had chosen to use one rootstock or own roots down the hill the differences in vine performance would have been much greater.

The benefits of using rootstocks to increase vine production uniformity

Rather than breaking blocks into smaller management units it will be possible to manage large vineyard blocks as a single unit regardless of the variation in soil or topography.

Things to consider

- Use soil surveys pre-planting (EM-38 and ground-truthing with soil pits)
- Requires an understanding of soil/scion/rootstock interactions
- Blocks should still be set up for differential management, with the long-term aim of managing as a single unit
- Matching rootstock to soil type down a row is best achieved if there is previous experience of rootstock use on the site

Scion and Rootstock Variety Vigour

Vine vigour has a major effect on final wine quality and it is well known that the best wine comes from balanced vines (Dry et al., 2004). Vine vigour is a product of a number of interactions, including the inherent vigour of the rootstock and scion variety. Table 10 and 11 lists the relative vigour levels of the common rootstock and scion varieties in South Australia and should be used in conjunction with the 'Choosing a Rootstock' section.

Table 10. The relative vigour levels of common and emerging scion varieties

Scion Variety	Relative Vigour
Shiraz	High
Sauvignon Blanc	High
Grenache	High
Semillon	High
Colombard	High
Mataro	High
Sultana	High
Tempranillo	High
Petit Verdot	High
Sangiovese	High
Chardonnay	Moderate
Cabernet Sauvignon	Moderate
Pinot Noir	Moderate
Verdelho	Moderate
Pinot Gris	Moderate
Riesling	Low moderate
Viognier	Low-moderate
Muscat Gordo	Low
Merlot	Low

Table 11. The relative vigour levels of common and emerging rootstock varieties

Rootstock Variety	Relative Vigour
Ramsey	High
140 Ruggeri	Moderate-High
1103 Paulsen	Moderate-High
99 Richter	Moderate-High
5BB Kober	Moderate-High
110 Richter	Moderate
SO4	Moderate
5C Teleki	Moderate
Schwarzmann	Low-Moderate
101-14	Low
420A	Low
3309C	Low

Regional Rootstock Recommendations

The aim of this section is to provide growers with information on the most appropriate rootstocks for their region with reference to the important varieties, viticultural issues and soil types of that region. As much as possible the information is based on results from commercial vineyards. Variety and rootstock planting data is taken from the PGIBSA vineyard register and is current as of May 2007.

Site Potential

In this section there is continual reference to site potential. Site potential describes the potential vigour (low, moderate or high) that will be conferred to a vine at a given site. Site potential is a combination of the following factors:

Potential Root-zone Depth

In the soil profile the depth to the impenetrable or impermeable layer is important as it determines the volume of soil that the vine roots can explore for moisture and nutrients. In general it can be said that the site potential increases with rooting depth.

Soil Fertility

The potential of a soil increases with soil fertility. Generally soil fertility increases with increasing soil clay content as clay soils have a greater capacity to store nutrients compared with sand (Gladstones, 1992).

Climate

Warm climates encourage growth so long as moisture is not a limiting factor. Cool to mild climates encourage growth because of the high rainfall and low evaporation that is generally associated with them. In general a warm, wet climate has greater potential for vigour than a cool, wet climate. In the absence of irrigation a cool, wet climate has greater potential for vigour than a warm, dry climate.

Regions

ADELAIDE HILLS

The Adelaide Hills is geographically a large area with a wide range of site-specific climatic attributes. The major varieties of the region are Chardonnay (26%, 986 ha), Sauvignon Blanc (20%, 777 ha) and Pinot Noir (14%, 543 ha). Cool and wet conditions during spring and late summer can negatively affect set and ripening leading to reductions in yield and quality. The irrigation water ranges from abundant supply and good quality in the higher rainfall areas to more marginal in the lower rainfall areas. Irrigation is normally required albeit at varying levels depending on the vineyard site. On the high potential sites of the Adelaide Hills it is often difficult to control vigour through water management, hence low vigour rootstocks should be a priority consideration at these sites. The soils of the typically higher rainfall areas of the Adelaide Hills are commonly highly leached and acidic in nature. Considerations for rootstock selection in the Adelaide Hills are therefore:

- Limiting vine vigour
- Increased fruitset
- Advanced ripening
- Waterlogging
- Soil acidity

Rootstock Use
Rootstock use in the Adelaide Hills accounts for 4.5% (172.3 ha) of total plantings. Schwarzmann has been the most widely planted rootstock in the last ten years (1997–2006) (Table 12)

Table 12. Rootstock plantings (ha) between 1997–2006 for the three major varieties of the Adelaide Hills.

ADELAIDE HILLS – Area Planted, in Hectares (1997-2006)								
Variety	101-14	5C Teleki	Schwarz-mann	Other r/stocks	Total r/stock	Ungrafted	Totals	% r/stock
Chardonnay	7	7	18	10	42	628	669	6.3
Pinot Noir	4	6	2	7	19	404	423	4.4
Sauvignon Blanc	1	1	7	7	17	622	639	2.6
Other varieties	4	8	12	53	76	1109	1186	6.5
Total	16	22	39	77	154	2763	2917	5.3

Recommended Rootstocks

101–14

101–14 is particularly well suited to the Adelaide Hills because of its positive influence on set, its ability to advance ripening (it generally ripens earlier than own roots by one week) and its ability to reduce vine vigour. 101–14 is the most widely used rootstock in New Zealand.

101–14 has proven quality outcomes in the Adelaide Hills with Chardonnay and Pinot Noir and has consistently produced premium quality fruit on moderate-high potential sites.

Data from a Pinot Noir/rootstock demonstration trial that compared the performance of 101–14, 5C Teleki and Schwarzmann with own roots indicated that 101–14 had a positive influence on the performance of Pinot Noir in terms of reduced vine vigour, increased wine colour and advanced ripening (Henschke, 2007).

The ability of 101–14 to reduce vine vigour has the potential to reduce canopy management costs and increase quality of Sauvignon Blanc on moderate-high potential sites.

This rootstock does not tolerate drought conditions so it may struggle on low potential sites in the warmer drier parts of the region (e.g. western foothills to the north of Gumeracha and the eastern foothills) or where water availability is not guaranteed. 101–14 is also susceptible to waterlogging in its early years.

5C Teleki

5C Teleki has a positive influence on set, may advance ripening and imparts moderate vigour to the scion.

5C Teleki has proven quality outcomes in the Adelaide Hills with Chardonnay and Pinot Noir and has consistently produced premium quality fruit on low-moderate potential sites. It is also well suited to Cabernet Sauvignon and Merlot because of its positive influence on set and ripening. The use of 5C Teleki with Sauvignon Blanc, except on low-moderate potential sites, is not advisable as it may result in excessive vigour.

5C Teleki is more tolerant of drought conditions than 101–14 and Schwarzmann so would be a better choice in the warmer drier parts of the region on shallow or restrictive soils or where water availability is not guaranteed.

Schwarzmann

Schwarzmann has been widely used in the Adelaide Hills because of its positive influence on set, its ability to advance ripening and its low-moderate vigour. Schwarzmann has proven quality outcomes in the Adelaide Hills with Chardonnay and has consistently produced premium quality fruit on moderate potential sites. Schwarzmann has been used successfully with Sauvignon Blanc on moderate-high potential sites in New Zealand.

Because of its low-moderate vigour Schwarzmann would also perform well with Pinot Noir; however, data from a Pinot Noir/rootstock demonstration trial indicated that Schwarzmann required greater acid addition than other rootstocks and own roots (Henschke, 2007). For this reason 101–14 may be a better low-moderate vigour rootstock choice.

Because it is susceptible to drought conditions this rootstock should not be used on shallow soils or where water availability is limited.

Other rootstocks

In most parts of the Adelaide Hills, 110 Richter, 1103 Paulsen and 140 Ruggeri would impart excessive vigour to the scion. However, they have been used at low potential sites in the warmer drier parts of the region around the western foothills to the north of Gumeracha and the eastern foothills near Mt Barker.

3309C imparts lower vigour to the scion than 101–14 and advances fruit maturity. It is a good rootstock for deep, well-drained cool soils that are well supplied with water and its use would probably be limited to high potential sites with Sauvignon Blanc and possibly Pinot Noir. It is unsuited to dry and shallow conditions and not appropriate for heavy soils (Cirami, 1999). As there is little experience in growing this variety in Australia a trial block would be advisable before undertaking a large-scale planting.

BAROSSA VALLEY

The Barossa Valley has a warm enough climate to fully ripen red wine varieties of premium quality. Shiraz is the major variety of the region and accounts for 48% (5041 ha) of total plantings. Irrigation water is generally in limited supply (water is generally applied at 1 ML/ha or less) and some underground sources are moderately saline. As a result, drought and salinity tolerance are major considerations in rootstock selection. There are areas of sandy soils where the presence of nematodes will impact on vine performance. A survey of 11 vineyards by PGIBSA found high numbers of root-knot nematode, citrus nematode and pin nematode. There are also areas of soil with high 'active lime' content and high soil pH. Therefore, the main considerations for rootstock selection in the Barossa Valley are likely to be:

- Drought tolerance/water-use efficiency
- Salinity tolerance
- Nematode resistance
- Soil lime content

Rootstock Use

Rootstock use in the Barossa Valley is relatively high, driven by the need for nematode resistant planting material, and accounts for 20% (1972 ha) of total plantings. The most widely planted rootstock is Ramsey; however 1103 Paulsen and 140 Ruggeri have been the preferred rootstocks over the last 10 years (Table 13).

Table 13. Rootstock plantings (ha) between 1997–2006 for the three major varieties of the Barossa Valley.

	BAROSSA VALLEY – Area Planted, in Hectares (1997-2006)							
Variety	1103 Paulsen	140 Ruggeri	Ramsey	Other r/stocks	Total r/stock	Ungrafted	Totals	% r/stock
Shiraz	181	139	54	182	556	2764	3319	20.1
Cabernet Sauvignon	58	23	74	49	204	579	784	35.2
Grenache	3	3	4	3	12	155	166	7.5
Other varieties	136	54	101	172	463	636	1099	72.7
Total	378	219	233	406	1235	4134	5368	29.9

Recommended Rootstocks

1103 Paulsen and 140 Ruggeri

Both rootstocks have excellent drought, salinity and lime tolerance. These rootstocks do impart moderate-high vigour to the scion so therefore, in order to avoid excessive vigour they should not be used on high potential sites.

On low-moderate potential sites (both sandy loams and red-brown earths) with low inputs of irrigation and fertiliser, these two rootstocks have consistently produced premium quality Shiraz and Cabernet Sauvignon.

140 Ruggeri has greater root-knot nematode tolerance than 1103 Paulsen so would be the preferred rootstock in nematode infested soils.

Observations from Victoria (Goulbourn Valley) have shown that these rootstocks are well suited to dry grown (minimal input) Grenache and Mataro. Because they increase vigour and productivity they are also well suited to the production of semi-premium to premium quality Chardonnay, Semillon and Riesling.

110 Richter

The use of 110 Richter in the Barossa Valley has been limited but it has excellent potential because it imparts moderate vigour to the scion variety (similar to ungrafted *V. vinifera)*. It also has good to excellent drought tolerance, greater tolerance of salinity than ungrafted vines and moderate root-knot nematode resistance. As 110 Richter is lower in vigour than both 140 Ruggeri and 1103 Paulsen it is the preferred option in situations with Shiraz on moderate-high potential sites.

101–14

101–14 is a low vigour rootstock with good salinity tolerance and high resistance to root-knot nematode. Research by PGIBSA has shown that it devigorates Shiraz and therefore this rootstock is particularly well suited to sites where vigour control is difficult. However because it is low vigour, avoid using this rootstock with low vigour scions (Merlot) and when planting white varieties on low-moderate potential sites. This rootstock is also susceptible to drought conditions so its use in the Barossa Valley should be limited to sites that have guaranteed water supplies.

Other rootstocks
Ramsey is drought-and salinity-tolerant and has broad nematode resistance and so would seem to be well suited to conditions in the Barossa Valley. However it is a high vigour rootstock and thus needs to be used with some caution. It has produced quality outcomes with Shiraz but quality can be inconsistent depending on the season and site (it does better in drought years and on low potential sites). It is a rootstock well suited to the production of semi-premium Chardonnay and Semillon.

5C Teleki and 5BB Kober have produced quality outcomes in the Barossa Valley but are susceptible to drought conditions and are moderately sensitive to saline conditions. Both will perform better in red-brown earths than sand-sandy loams.

Schwarzmann has very similar characteristics to 101–14 and has produced quality outcomes in the Barossa Valley; however, because it accumulates potassium at high levels, 101–14 is the preferred low vigour rootstock.

CLARE VALLEY

The Clare Valley has a climate warm enough to fully ripen red wine varieties of premium quality. Shiraz and Cabernet Sauvignon account for 33% (1895 ha) and 21% (1256 ha) of total plantings respectively. The region is also noted for its premium white varieties, particularly Riesling, which accounts for 21% (1236 ha) of total area. Most vineyards require irrigation, which comes predominately from the aquifer and dams and can be reasonably saline, and in limited supply. Water is generally applied at 1 ML/ha or less. Those vineyards not utilising underground water generally pay for supplementary water on a per volume basis. As a result water-use efficiency is a major consideration in rootstock selection. The main considerations for rootstock selection in the Clare Valley are likely to be:

- Drought tolerance/water-use efficiency
- Salinity tolerance
- Soil lime content

Rootstock Use

Less than 3% (147 ha) of the Clare Valley is planted on rootstock; Table 14 shows that between 1997 and 2006 there has been a strong preference for Riesling on 1103 Paulsen and 140 Ruggeri on Cabernet Sauvignon.

Table 14. Rootstock plantings (ha) between 1997–2006 for the three major varieties of the Clare Valley.

Variety	1103 Paulsen	110 Richter	140 Ruggeri	Other r/stocks	Total r/stock	Ungrafted	Totals	% r/stock
Shiraz	0	6	0	14	20	1212	1232	1.6
Cabernet Sauvignon	0	0	12	7	19	725	744	2.6
Riesling	20	2	0	9	31	715	746	4.2
Other varieties	0	0	6	39	46	561	607	7.5
Total	20	8	18	69	116	3213	3329	3.5

CLARE VALLEY – Area Planted, in Hectares (1997-2006)

Recommended Rootstocks

1103 Paulsen and 140 Ruggeri

Both rootstocks have excellent drought, salinity and lime tolerance but because they impart moderate-high vigour to the scion they may not be appropriate across the full range of soil types and varieties in the Clare Valley.

Both rootstocks hold onto their leaves and maintain canopy 'freshness' under stress more readily than other rootstocks and own roots and so they are well suited to the production of Riesling on low-moderate potential sites, sites prone to salinity or where water availability is not always guaranteed. For the production of premium quality Shiraz and Cabernet Sauvignon they should be used at low and moderate potential sites and be managed with reduced water and fertiliser inputs.

110 Richter

110 Richter has good-excellent drought tolerance, greater tolerance of salinity than ungrafted vines and some lime tolerance. It imparts lower vigour to the scion compared with 140 Ruggeri and 1103 Paulsen so is a good option on the moderate-high potential sites of the region.

101–14

101–14 is a low vigour rootstock with good salinity tolerance and good resistance to root-knot nematode. Research by PGIBSA has shown that it devigorates *Vitis vinifera* and so this rootstock is particularly well suited to sites where controlling vigour is difficult (e.g. Shiraz on high potential sites). However because, it is low vigour, avoid using this rootstock with Riesling and Merlot on low-moderate potential sites. 101–14 performs poorly in drought conditions so should only be used where the supply of water is guaranteed.

COONAWARRA

Coonawarra is a cool climate region capable of producing a range of wine styles including premium quality medium to full bodied red wines. Cabernet Sauvignon is the major variety of the region and accounts for 58% (3415 ha) of total area. Vine vigour can be naturally high, particularly on the black cracking clays, and the selection of moderate vigour rootstocks may therefore be appropriate. There is always a risk of inclement weather late spring and during ripening. Irrigation is used in most vineyards and is taken from the aquifer, which has rising salinity levels. The region is moving from area-based licences to a volumetric licence in line with the national water initiative. Water-use reductions as high as 30% may be possible (P. Balnaves, pers. comm.), therefore water-use efficiency has become an important consideration. The soils of the region generally have high lime content. The main considerations for rootstock selection at Coonawarra are therefore likely to be:

- Vigour reduction
- Drought tolerance/water-use efficiency
- Soil lime content
- Increased set
- Advanced ripening

Rootstock Use

Rootstock use in Coonawarra is currently less than 2% (85 ha). 5C Teleki has been the most widely planted rootstock in the last ten years (1997–2006). (Table 15)

Table 15. Rootstock plantings (ha) between 1997–2006 for the three major varieties of the Coonawarra.

COONAWARRA – Area Planted, in Hectares (1997-2006)							
Variety	5C Teleki	5BB Kober	Schwarz-mann	Other r/stocks	Total r/stock	Ungrafted	Totals	% r/stock
Cabernet Sauvignon	12	3	0	8	23	1566	1588	1.4
Shiraz	0	16	0	3	19	406	425	4.6
Merlot	4	0.5	7	1	12	237	250	4.9
Other varieties	0.5	0	0	11	12	170	182	6.3
Total	16.5	19.5	7	23	66	2379	2445	2.7

Recommended Rootstocks

5C Teleki

5C Teleki is particularly suited to this region because of its positive influence on set, its ability to advance ripening (it ripens earlier than 5BB Kober) and its moderate vigour. It also has good tolerance of calcareous soils. 5C Teleki has proven quality outcomes in Coonawarra with Cabernet Sauvignon and has consistently produced super premium quality fruit. 5C Teleki has also performed well with Merlot because it slightly invigorates the scion and has a positive influence on fruitset. 5C Teleki would also perform well with Shiraz on low-moderate potential sites, based on observations from Wrattonbully.

5BB Kober

This rootstock has similar characteristics to 5C Teleki but imparts higher vigour to the scion and thus is better suited to the shallower, less fertile soils. It has proven quality outcomes in this region and has consistently produced super premium quality Cabernet Sauvignon. 5BB Kober has better drought tolerance than 5C Teleki. Excessive vigour may result if 5BB Kober is grafted to Shiraz and planted on moderate or high potential sites.

99 Richter and 110 Richter

99 Richter has been used with success in Coonawarra and it would be expected that 110 Richter would also perform well. Both have good drought tolerance (110 Richter has lower vigour and better drought tolerance than 99 Richter) and similar tolerance to saline conditions as 5C Teleki and 5BB Kober. These rootstocks would therefore be a good choice at sites where water availability is not guaranteed. Both rootstocks tolerate lime.

Schwarzmann

Schwarzmann has been used successfully on high potential sites with Shiraz in Coonawarra and would be expected to perform well with Cabernet Sauvignon. It has low tolerance of lime which could be an issue, but does tolerate waterlogging better than most rootstocks. It is moderately tolerant of saline conditions but susceptible to drought so should only be used at sites with adequate water supplies.

101–14

101–14 is yet to be used at Coonawarra but should be considered because of its low vigour, positive influence on fruitset and ability to advance ripening. While it is moderately tolerant of lime (greater than Schwarzmann) and saline conditions it is susceptible to drought so should only be used on deeper soils where water is available on demand. This rootstock is susceptible to waterlogging in its early years.

Other rootstocks
420A is a low vigour rootstock with higher lime tolerance than 101–14 and Schwarzmann. It would therefore be a good alternative to these rootstocks in the high lime soils of the Coonawarra. It must be noted; however, that 420A performs poorly in drought and saline conditions.

EDEN VALLEY

Eden Valley is a cool to warm climate region capable of producing aromatic white wines through to medium to full bodied red wines depending on vineyard site. Shiraz (29%, 650 ha), Riesling (25%, 551 ha) and Cabernet Sauvignon (14%, 335 ha) are the principal varieties of the region. Eden Valley is largely dependent on irrigation sourced from water harvesting in winter. Vine water-use efficiency is therefore a major consideration in rootstock selection. Eden Valley is generally between 400 m and 600 m above sea level, with most grapegrowing country located in the cool, wet parts of the region. In these parts of the region there are generally shallow, sandy soils that are moderately acidic and prone to spring waterlogging. Further to the north and east of Eden Valley, the rainfall diminishes and soils can become neutral to alkaline. Therefore the main considerations for rootstock selection in Eden Valley are likely to be:

- Drought tolerance/water-use efficiency
- Waterlogging
- Soil acidity
- Fruitset
- Soil lime content (north-east Eden Valley).

Rootstock Use

Rootstock use accounts for less than 13% (293 ha) of plantings in Eden Valley. The most commonly used rootstock in the last 10 years (1997–2006) is 1103 Paulsen (Table 16).

Table 16. Rootstock plantings (ha) between 1997–2006 for the three major varieties of Eden Valley.

EDEN VALLEY – Area Planted, in Hectares (1997-2006)							
Variety	1103 Paulsen	Schwarz-mann	140 Ruggeri	Other r/stocks	Total r/stock	Ungrafted	Totals	% r/stock
Shiraz	9	0	2	12	23	288	319	7.2
Riesling	38	17	2	34	90	114	373	24.1
Cabernet Sauvignon	4	1	10	0	15	208	246	6.1
Other varieties	23	0.5	18	44	86	214	425	20.1
Total	74	18.5	32	90	214	824	1363	15.7

Recommended Rootstocks

1103 Paulsen

1103 Paulsen has excellent drought and salinity tolerance and also tolerates acid soils. 1103 Paulsen has performed well with Riesling and growers report that it has out-performed 101–14 and 5C Teleki as it holds onto its leaves until later in the season. It has also performed well with both Viognier and Cabernet Sauvignon.

140 Ruggeri

140 Ruggeri has excellent drought and salinity tolerance and also tolerates acid soils. It can produce moderate-high vigour when grafted to Riesling, Shiraz, Cabernet Sauvignon and Merlot with limited inputs on the thick loamy sand over brown sandy clay loam in the less elevated area of Eden Valley. Growers should avoid using 140 Ruggeri on high potential sites with high vigour scion varieties.

110 Richter

This rootstock has good-excellent drought tolerance and greater tolerance of salinity than ungrafted vines. 110 Richter is lower in vigour than both 140 Ruggeri and 1103 Paulsen and it would be the preferred option in high potential sites. It has not been widely used with Riesling, but one would expect it to perform similarly to ungrafted Riesling, with the added benefit of drought tolerance.

5BB Kober

5BB Kober has been used successfully for premium Riesling production in High Eden on a vigour-limiting shallow soil. 5BB Kober is only moderately tolerant of drought conditions, so use of this rootstock should be avoided where water is likely to be limited.

101–14

101–14 is suited to high vigour scions and or the high potential sites of the more elevated (colder) parts of Eden Valley, as long as there is water available. In the warm, dry parts of the region this rootstock can struggle late in the season with premature leaf loss, particularly in sandy or shallow soils.

LANGHORNE CREEK

Langhorne Creek has a climate capable of producing semi-premium and premium wine styles.

Plantings in the region are dominated by Shiraz (33%, 2096 ha), Cabernet Sauvignon (32%, 2053 ha) and Chardonnay (12%, 811 ha). Vines are generally moderate in vigour, but selection of the most appropriate rootstock will depend largely on the required end use of the grapes produced.

Ninety-five percent of vineyards are irrigated from the nearby Lake Alexandrina (Dry et al., 2004). There has been a concerted effort to reduce irrigation from an environmental perspective and as a result vine water-use efficiency is an important consideration. Soil and water salinity is also increasingly becoming an important issue in Langhorne Creek. High populations of root-knot, citrus and root-lesion nematodes have been reported in the region. The main considerations for rootstock use are therefore:

- Salinity tolerance
- Drought tolerance/water-use efficiency
- Nematode resistance

Rootstock Use

Rootstock use in Langhorne Creek accounts for 15% (934 ha) of total plantings. However, in response to the issues outlined above, grafted vines have accounted for 33% (346 ha) of new plantings since 2002. The most popular rootstocks in the last ten years have been Schwarzmann, 5C Teleki and 1103 Paulsen (Table 17).

Table 17. Rootstock plantings (ha) between 1997–2006 for the three major varieties of Langhorne Creek.

LANGHORNE CREEK – Area Planted, in Hectares (1997-2006)							
Variety	Schwarz-mann	1103 Paulsen	5C Teleki	Other r/stocks	Total r/stock	Ungrafted	Totals	% r/stock
Shiraz	41	36	23	47	146	1274	1420	10.3
Cabernet Sauvignon	42	28	32	100	200	876	1076	18.6
Chardonnay	4	3	53	78	139	426	565	24.6
Others	81	12	28	101	222	797	1018	21.8
Total	168	79	136	326	707	3373	4079	17.3

Recommended Rootstocks

5C Teleki

5C Teleki imparts moderate vigour to the scion and is suited to the production of semi-premium and premium Cabernet Sauvignon, Shiraz and Chardonnay. In a trial by Gawel et al. (2000) that compared the sensory properties of Cabernet Sauvignon grafted to 4 rootstocks (Ramsey, 110 Richter, Schwarzmann and 5C Teleki) and own roots, the 5C Teleki treatment was described as 'the best of them'. 5C Teleki is moderately susceptible to both saline and drought conditions, so before using this rootstock growers need to ensure that they have adequate supplies of good quality water.

110 Richter

110 Richter has good-excellent drought tolerance, moderate root-knot nematode resistance and greater tolerance of salinity than ungrafted vines. It imparts lower vigour to the scion compared with 140 Ruggeri and 1103 Paulsen so is a good option on the moderate-high potential sites of the region (flood plains). Growers report that on these high potential sites, it has performed similarly to ungrafted Shiraz, Cabernet Sauvignon and Chardonnay, but with lower water inputs (consistently managed with less than 1 ML/ha).

140 Ruggeri and 1103 Paulsen

Both have excellent drought and salinity tolerance; however, because they impart moderate-high vigour to the scion they may not be appropriate for premium production across the full range of soil types of Langhorne Creek. Because they increase productivity they are well suited to the production of semi-premium quality wine grapes for the major varieties of the region. However, with appropriate selection and management they can also produce premium quality. For quality purposes they should be used in shallow, low fertility soils and be managed with reduced inputs.

101–14

101–14 is a low vigour rootstock with good salinity tolerance and good resistance to root-knot nematode. Research by PGIBSA has shown that it devigorates *Vitis vinifera* and so this rootstock is particularly well suited to sites where vigour control is difficult (deep, alluvial soils). However because it imparts low vigour avoid using this rootstock with low vigour scions (e.g. Merlot) and when planting white varieties on low-moderate potential sites. This rootstock is also susceptible to drought conditions so it should not be used on sites with limited water supplies.

Other rootstocks

Schwarzmann is the most widely planted rootstock in the region. However, its popularity has reduced over the last five years in favour of 5C Teleki and the drought-tolerant rootstocks 1103 Paulsen and 110 Richter. Schwarzmann has produced premium Cabernet Sauvignon on very thick sand over red clay and would also perform well on deep, alluvial soils because it generally imparts low vigour to the scion. It is moderately tolerant of saline conditions and has good nematode resistance so would seemingly be well suited to the region. However its high potassium uptake, susceptibility to drought conditions, and isolated instances in Langhorne Creek relating to the poor performance and death of Cabernet Sauvignon and Merlot vines grafted to this rootstock, should probably lead growers to use other rootstocks.

McLAREN VALE

McLaren Vale has a climate capable of producing a range of premium wine styles. The two major varieties of the region are Shiraz (48%, 3356 ha) and Cabernet Sauvignon (19%, 1328 ha). Most vineyards require irrigation. While the use of recycled water now accounts for up to 15% of the irrigated area in McLaren Vale, 70% of vineyards are irrigated from underground sources which can vary in quantity and quality (Dry et al., 2004). Salinity is a problem particularly on the restrictive duplex soils of the region. Water stress is another key issue particularly on shallow soils. The use of ungrafted vines in the sandy soils of Blewitt Springs may not be sustainable due to high nematode numbers. The main considerations for rootstock use in McLaren Vale are therefore:

- water-use efficiency/drought tolerance
- Salinity tolerance
- Nematode resistance

Rootstock Use

Total rootstock use in McLaren Vale accounts for less than 8% (543 ha) of total plantings, however since 2002, 20% (117 ha) of new plantings have been on rootstock. The drought- and salinity-tolerant rootstocks, 1103 Paulsen, 140 Ruggeri and Ramsey have been the most widely planted rootstocks in the last ten years (Table 18).

Table 18. Rootstock plantings (ha) between 1997–2006 for the three major varieties of McLaren Vale.

McLAREN VALE – Area Planted, in Hectares (1997-2006)							
Variety	1103 Paulsen	Ramsey	140 Ruggeri	Other r/stocks	Total r/stock	Ungrafted	Totals	% r/stock
Shiraz	53	13	14	86	165	1813	1978	8.4
Cabernet Sauvignon	14	3	7	6	31	460	491	6.3
Chardonnay	20	5	0	24	49	128	177	27.7
Other varieties	26	21	13	42	101	519	620	16.3
Total	113	42	34	158	346	2920	3266	10.6

Recommended Rootstocks

1103 Paulsen and 140 Ruggeri

Both have excellent drought and salinity tolerance. 140 Ruggeri has good resistance and 1103 Paulsen has moderate resistance to root-knot nematode. Both rootstocks have been used with success across a range of varieties and soil types, although some care needs to be taken in situations combining high vigour scions (e.g. Shiraz) on moderate-high potential sites. 140 Ruggeri has greater root-knot nematode resistance than 1103 Paulsen so would be the preferred rootstock in nematode infested soils.

These rootstocks perform best with Shiraz on low-moderate fertility soils at a number of sites in McLaren Vale including ironstone-rich loamy sand over red mottled clay and deep bleached sands. While 140 Ruggeri and 1103 Paulsen have not been used extensively with Cabernet Sauvignon in McLaren Vale one would expect both of these rootstocks to be well suited to this variety. In Victoria (Goulburn Valley) these rootstocks have been used for dry grown/minimal input Grenache. Both would also be well suited to the production of semi-premium to premium Chardonnay and Riesling.

110 Richter

110 Richter has good-excellent drought tolerance, greater tolerance of salinity than ungrafted vines and some lime tolerance. Its use in McLaren Vale has been limited but because it imparts lower vigour to the scion compared with 140 Ruggeri and 1103 Paulsen it would be a good option on the moderate-high potential sites of the region.

101–14

101–14 is a low vigour rootstock with good salinity tolerance and good resistance to root-knot nematode. Research by PGIBSA has shown that it devigorates Shiraz and so this rootstock is particularly well suited to sites where vigour control is difficult. However, because it imparts low vigour, avoid using this rootstock with low vigour scions (e.g. Merlot) and when planting white varieties on low-moderate potential sites. This rootstock is also susceptible to drought conditions so should not be used at sites with limited water supplies.

Other rootstocks
Ramsey has excellent salinity and drought tolerance and high nematode resistance, but does impart high vigour to the scion, so some care needs to be taken with the selection and management of this rootstock. This rootstock has been used with success with a number of varieties, particularly when growers have used low water and fertiliser inputs on the low fertility sands of Blewitt Springs.

PADTHAWAY

Padthaway has a climate capable of producing a range of premium wine styles. The principal varieties of the region are Shiraz (29%, 1171 ha), Chardonnay (24%, 1109 ha) and Cabernet Sauvignon (20%, 821 ha). Vine vigour is generally moderate to high and the selection of moderate vigour rootstocks may be appropriate. Irrigation is used in most vineyards and water is taken from the aquifer, which varies in salinity from site to site. Diligent use of water is required due to environmental issues; hence water-use efficiency is an important consideration in rootstock selection. There are also areas of sandy soils where the presence of nematodes may impact on vine performance. The main considerations for rootstock selection at Padthaway are therefore likely to be:

- Drought tolerance/water-use efficiency
- Salinity
- Vigour reduction
- Soil lime content
- Nematode resistance

Rootstock Use

Rootstock use at Padthaway is currently less than 9% (329 ha); however since 2002, 25% (166 ha) of plantings have been on rootstocks. This is largely a response to the need for vines with a degree of salinity and drought tolerance. 1103 Paulsen has been the most widely planted rootstock in the last ten years (1997–2006) (Table 19).

Table 19. Rootstock plantings (ha) between 1997–2006 for the three major varieties of Padthaway.

PADTHAWAY – Area Planted, in Hectares (1997-2006)							
Variety	1103 Paulsen	140 Ruggeri	5C Teleki	Other r/stocks	Total r/stock	Ungrafted	Totals	% r/stock
Shiraz	15	0	8	29	52	577	629	8.3
Chardonnay	39	11	6	57	113	288	401	28.1
Cabernet Sauvignon	0	0	6	0	6	367	373	1.6
Other varieties	32	0	11	28	71	151	221	32.0
Total	86	11	31	114	242	1383	1624	14.9

Recommended Rootstocks

1103 Paulsen and 140 Ruggeri

Both have excellent drought and salinity tolerance; however, because they impart moderate to high vigour to the scion they may not be appropriate across the full range of soil types. With appropriate selection and management they can produce premium quality. For quality purposes they should be used in shallow, low fertility soils and be managed with reduced inputs. Because they increase productivity they are also well suited to the production of semi-premium quality wine grapes for the major varieties of the region.

5C Teleki

5C Teleki is suited to this region because of its positive influence on set, its ability to advance ripening and its moderate vigour. It also has good tolerance of calcareous soils. 5C Teleki has proven quality outcomes in Coonawarra and Wrattonbully with Cabernet Sauvignon and would perform well with Cabernet Sauvignon and Chardonnay on the moderate-high potential sites and with Shiraz on the moderate potential sites of Padthaway. 5C Teleki is classified as moderately susceptible to saline conditions, but it is less susceptible to salinity than ungrafted *Vitis vinifera*.

110 Richter

110 Richter has good to excellent drought tolerance, greater tolerance of salinity than ungrafted vines and some lime tolerance. Its use in Padthaway has been limited; however, because it imparts lower vigour to the scion compared with 140 Ruggeri and 1103 Paulsen it would be a good option on the moderate-high potential sites of the region.

101–14

101–14 is a low vigour rootstock with good salinity tolerance. Research by PGIBSA has shown that it devigorates *Vitis vinifera* and so this rootstock is well suited to sites where vigour control is difficult. However, because it imparts low vigour, avoid using this rootstock with low vigour scions such as Merlot and when planting white varieties (Chardonnay) on low-moderate potential sites as the resulting vines may be under-vigorous. This rootstock is also susceptible to drought conditions so should not be used on sites with limited water supplies.

RIVERLAND

The Riverland is a hot climate region that produces predominantly commercial quality wine grapes. Shiraz (24%, 5478 ha), Chardonnay (21%, 4864 ha) and Cabernet Sauvignon (15%, 3448 ha) are the most widely planted varieties of the region. In most cases when inputs are administered correctly vine vigour can be reasonably high which is suitable for the end use requirements of the grapes.

Licensed water is used from the River Murray with annual licence and volume usage payments. Allocations have been reduced in recent years in response to the drought conditions. As a result, vine water-use efficiency and drought tolerance are extremely important considerations.

The salinity of water from the River Murray can fluctuate and there are issues in relation to perched water tables that are highly saline in parts of this region, so the use of salinity-tolerant rootstocks should be strongly considered.

Lime tolerance is another important consideration as there are a large percentage of calcareous soils utilised in the Riverland. Nematodes, particularly root-knot nematode are a major consideration on the sandy soils of the region. The main considerations for rootstock selection are therefore:

- Drought tolerance/water-use efficiency
- Salinity tolerance
- Nematode resistance
- Soil lime content

Rootstock Use

Rootstock use in the Riverland accounts for 40% (8886 ha) of total area and since 2002, 75% (3338 ha) of plantings have been on rootstock. Ramsey has been the most widely planted rootstock in the last ten years (1997–2006) accounting for 30% (1715 ha) of total rootstock plantings, followed by 1103 Paulsen (27% or 1512 ha) and 140 Ruggeri (13% or 753 ha) (Table 20).

Table 20. Rootstock plantings (ha) between 1997–2006 for the three major varieties of the Riverland

RIVERLAND – Area Planted, in Hectares (1997-2006)							
Variety	Ramsey	1103 Paulsen	140 Ruggeri	Other r/stocks	Total r/stock	Ungrafted	Totals	% r/stock
Shiraz	90	217	106	445	857	2063	2920	29.4
Chardonnay	899	643	215	223	1980	829	2810	70.5
Cabernet Sauvignon	108	115	136	407	766	1454	2220	34.5
Other varieties	618	538	297	633	2085	1903	3989	52.3
Total	1715	1513	753	1708	5688	6249	11939	47.7

Recommended Rootstocks

Ramsey

Ramsey has excellent drought and salinity tolerance, broad nematode resistance and promotes yield thereby making it an ideal rootstock for the Riverland. It does impart high vigour to the scion so some care needs to be taken in terms of managing water and fertiliser inputs, particularly when grafting to high vigour scions such as Shiraz. There have been issues with Ramsey negatively effecting quality, however this has occurred mainly when growers have managed Ramsey in the same manner as *Vitis vinifera*. It may not be as suited to very heavy, wet soils or where underground water tables are present. Experience has shown that Ramsey performs best when water and fertiliser inputs are kept to a minimum.

1103 Paulsen and 140 Ruggeri

Both 1103 Paulsen and 140 Ruggeri have excellent drought, salinity and lime tolerance, making them particularly well suited to the region. They impart moderate-high vigour to the scion (less than Ramsey) and they consistently out-yield own-rooted vines. 140 Ruggeri has a higher tolerance of calcareous soils than 1103 Paulsen Both are well suited to all varieties and soil types, although 1103 Paulsen has performed poorly at some sites with high root-knot nematode populations.

110 Richter

110 Richter has not been widely planted in the Riverland. It has struggled to produce adequate vigour at sandy sites but has performed well with Shiraz on the heavy clay soils of the region. Some observations indicate high susceptibility to citrus nematode.

101–14

101–14 performs well in heavier textured soils and it is an excellent rootstock for controlling/ reducing vigour and yield. If managed correctly it can produce excellent grape quality. It does not perform well in infertile soils low in organic matter or in some replant situations. It has low resistance to citrus and root-lesion nematode. Considering its high water requirements and the possibility of future water restrictions in this region, it may not be the best long-term choice.

WRATTONBULLY

Wrattonbully is a cool climate region capable of producing medium to full bodied premium red wines. Plantings in the region are dominated by Cabernet Sauvignon (47%, 1205 ha) and Shiraz (26%, 681 ha). There is always a risk of inclement weather in late spring, which can affect fruitset particularly with the more susceptible varieties such as Merlot and Cabernet Sauvignon.

Irrigation is used in all vineyards and is taken from the aquifer, which varies in salinity from site to site. Most vineyards have a licensed water allocation with no usage-based payments required and so there is generally adequate water for the maintenance of the vine canopy throughout the growing season. There is a large degree of variation in the soil across the region ranging from deep sandy loams to shallow red clays of calcareous rock.

The main considerations for rootstock selection in Wrattonbully are:

- Salinity tolerance
- Increased fruitset
- Advanced ripening

Rootstock Use
Rootstock use in Wrattonbully is currently 8% (206 ha). Grafted Merlot accounts for 25% (45 ha) of the rootstock use. Rootstock use with the major variety of the region, Cabernet Sauvignon, accounts for less than 3% (22 ha). 5C Teleki, 1103 Paulsen and

101–14 have been the most widely planted rootstocks in the last ten years (1997–2006) (Table 21).

Table 21. Rootstock plantings (ha) between 1997–2006 for the three major varieties of Wrattonbully.

WRATTONBULLY – Area Planted, in Hectares (1997-2006)							
Variety	5C Teleki	1103 Paulsen	101–14	Other r/stocks	Total r/stock	Ungrafted	Totals	% r/stock
Shiraz	0	5	6	9	20	516	536	3.7
Cabernet Sauvignon	1	14	2	5	22	791	813	2.7
Merlot	14	2	2	27	46	225	271	16.9
Other varieties	11	9	5	59	84	166	250	33.6
Total	26	30	15	100	172	1698	1870	9.2

Recommended Rootstocks

5C Teleki

5C Teleki is suited to the region because of its moderate vigour, positive influence on fruitset, ability to advance ripening (ripens earlier than 5BB Kober) and because it has some tolerance of saline conditions and calcareous soils. 5C Teleki is the most widely planted rootstock as it performs well across a range of soil types and varieties. It has proven quality outcomes with Cabernet Sauvignon and Shiraz. 5C Teleki is also suited to Merlot because of its positive influence on set.

5BB Kober

5BB Kober has similar characteristics to 5C Teleki but imparts higher vigour to the scion and has greater drought tolerance and so is better suited to the shallower soils of the region. It has proven quality outcomes with Cabernet Sauvignon, Shiraz and Chardonnay in Wrattonbully. 5BB Kober is classified as moderately susceptible to saline conditions, but is less susceptible to salinity than ungrafted *Vitis vinifera*. It may promote high vigour when grafted to Shiraz on moderate-high potential sites.

140 Ruggeri and 1103 Paulsen

Both have excellent drought and salinity tolerance but because they impart moderate high vigour to the scion they may not be appropriate across the full range of soil types and varieties in Wrattonbully. For Merlot and white varieties where some invigoration of the scion is beneficial, these rootstocks would be appropriate. 140 Ruggeri has proven quality outcomes with Chardonnay. For the production of premium quality Shiraz and Cabernet Sauvignon they should be used in shallower, lower fertility soils and be managed with lower inputs.

Schwarzmann

Schwarzmann is a low-moderate vigour rootstock that has produced quality outcomes with Shiraz and Cabernet Sauvignon on the moderate-high potential sites of the region. It is not as well suited to the production of premium Chardonnay as it does not maintain the late season canopy freshness as well as 5BB Kober and 140 Ruggeri. This rootstock is susceptible to drought so it should only be used where water availability is guaranteed.

Other Rootstocks

110 Richter has good-excellent drought tolerance and greater tolerance of salinity than ungrafted vines. It imparts lower vigour to the scion compared with 140 Ruggeri and 1103 Paulsen so is a good option for Shiraz and Cabernet Sauvignon on the moderate-high potential sites of the region.

101–14 is a low vigour rootstock with good salinity tolerance. Research by PGIBSA has shown that it devigorates *Vitis vinifera* and so this rootstock is particularly well suited to sites where vigour control is difficult. However, because it imparts low vigour, avoid using this rootstock with low vigour scion varieties (e.g. Merlot) and when planting white varieties on low-moderate potential sites. This rootstock is also susceptible to drought conditions so it should not be used on sites with limited water supplies.

Choosing a Rootstock

With over a dozen rootstocks to choose from and an assortment of site factors affecting rootstock performance, choosing a rootstock can be a complicated exercise. The following section aims to simplify the process.

Step 1. Understand the rootstocks

Read the rootstock characteristics section to learn about the unique characteristics of the commonly used rootstocks in South Australia or for a quick reference, use Table 22 which is a summary of the rootstock characteristics. Also read the relevant regional recommendations section to identify which rootstocks have performed best in your region.

Table 22. A summary of the characteristics of the common rootstocks

Group	Rootstock	Relative Vigour	Nematode Resistance (Root-Knot)	Lime Tolerance	Salinity Tolerance	Drought Tolerance	Soil Acidity	Water-logging
A	101-14	L	H	▶	MT	S	○	○
	Schwarzmann	L-M	H	▶	MT	S	○	▶
	3309C	L	L	▶	S	S	○	○
B	SO4	M	MH	●	MS	MS	○	▶
	5C Teleki	M	M	●	MS	MS	○	○
	5BB Kober	M-H	H	●	MS	MS	○	○
	420A	L	M	●	S	S	○	○
C	110 Richter	M	M	●	MS	T	▶	▶
	1103 Paulsen	M-H	MH	●	MT	T	▶	▶
	99 Richter	M-H	MH	●	MT	T	▶	○
	140 Ruggeri	M-H	H	●	MT	HT	●	○
D	Ramsey	H	H	▶	MT	HT	○	○
E	*Vitis Vinifera*	M-L	○	●	S	MS	▶	▶

Key to table:

● Good   ▶ Moderate   ○ Poor   L = Low   M = Medium   H = High

S = Susceptible MS = Moderately susceptible   MT = Moderately tolerant

T = Tolerant   HT = Highly Tolerant

Step 2. **Create a balanced vine**

Rootstocks influence vine vigour and therefore vine balance. It is a widely accepted that the best quality wines are produced from balanced vines, so the most important selection criterion is to choose a rootstock that contributes to the production of a balanced vine. There are three parts to this step.

Match the rootstock to the site potential and scion variety

Doing this requires an understanding of the vigour level of your scion variety (see page 41) and the potential of your site. Site potential is a function of soil depth, soil fertility and the climate. For further explanation on site potential see page 43.

Below are five examples from commercial vineyards (unless otherwise indicated) that illustrate the interactions between the potential of the site and the inherent vigour of the scion and rootstock.

In the following example (see next page for illustrations), the combination of a high vigour site, a high vigour scion and a high vigour rootstock has led to the production of a high vigour, out-of-balance vineyard. A more appropriate rootstock in this case would have been a low or moderate vigour rootstock, possibly 101–14 (if water was available) or 110 Richter.

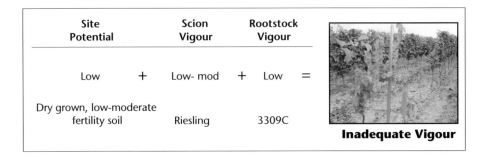

Site Potential		Scion Vigour		Rootstock Vigour	
Moderate-high	+	High	+	High	=
Thick sand over red mottled clay		Shiraz		140 Ruggeri	

Excessive Vigour

In the next example, which is taken from a replicated rootstock trial, the combination of a low potential site, coupled with a low-moderate vigour scion and rootstock has produced vines with inadequate vigour. At this site a higher vigour rootstock such as 110 Richter or 5C Teleki would have been more appropriate.

Site Potential		Scion Vigour		Rootstock Vigour	
Low	+	Low- mod	+	Low	=
Dry grown, low-moderate fertility soil		Riesling		3309C	

Inadequate Vigour

In the next three examples (facing page) the rootstock has been matched to the potential of the site and the inherent vigour of the scion leading to the production of balanced vines and quality outcomes.

Growers can use this method to determine the rootstock vigour (high, medium or low) and therefore the range of rootstocks that would contribute to vine balance at their site. The relative vigour levels of the rootstocks can be found on page 42.

Consider the end-product objectives

In order to create a balanced vine one should consider the end-product objectives. Quality may be less important than yield, or vice versa. Higher vigour (capacity) rootstocks (Ramsey, 140 Ruggeri and 1103 Paulsen) are better suited to the production of commercial grade wine-grapes because they produce canopies that will be in balance with the higher yields. Some rootstocks are better suited to low yields and high quality e.g. 101–14. However, it is important to note that high vigour rootstocks can produce high quality wine grapes so long as they are suitably managed

Site Potential		Scion Vigour		Rootstock Vigour		
Moderate	+	Moderate	+	Moderate	=	
Dark clay loam over rubbly calcrete		Cabernet Sauvignon		5C Teleki		**Balanced vine**

Site Potential		Scion Vigour		Rootstock Vigour		
Low	+	Low-moderate	+	Moderate-high	=	
Shallow sandy loam over basement rock		Riesling		5BB Kober		**Balanced vine**

Site Potential		Scion Vigour		Rootstock Vigour		
Moderate-High	+	High	+	Low	=	
Loam over red clay on weathering rock		Shiraz		101–14		**Balanced vine**

and matched with an appropriate soil type and climate and that the use of a low vigour rootstock does not always lead to quality wine grape production. Growers should also be aware that in warm-hot climates some varieties (particularly white varieties) may benefit from slightly higher vigour rootstocks to reduce bunch exposure and sun damage. In cool climates a slightly lower vigour rootstock maybe beneficial to increase bunch exposure.

Moderate vigour by reducing inputs (optional)

While it is important to finish with a balanced vine some vineyard managers may choose to use a rootstock that, under a normal production system, would produce more than the desired vigour levels. This is done with the knowledge that if they manage the vineyard with reduced inputs (water and fertiliser) it will decrease vine vigour, produce a balanced vine and potentially lead to greater profitability through the reduction in the cost of production. This is best attempted in regions where irrigation is a necessity i.e those regions with a winter dominant rainfall.

Step 3. Consider the site issues

Once you have established which rootstocks will contribute to vine balance, read the Site Factors section (page 24) to identify the important issues at your site, then modify your rootstock selection accordingly.

Step 4. Make your Final Decision

When making your final decision consider what has performed well on the site in the past or what is currently performing well in the area. Discuss your final selections with fellow growers, your winery and nursery. It may be appropriate to use 2–3 different rootstocks depending on the soil type and topography (see page 39).

The successful use of a rootstock does not end with the final selection; appropriate management strategies should also be considered. The rootstock management section will give you an insight into how to best manage individual rootstocks.

Additional Considerations in using Rootstocks

Purchasing Grafted Planting Material

More forward planning is required when using rootstocks than ungrafted vines because the plants need to be grafted and then spend time in the field during the growing season. It is often not possible to source a particular rootstock-scion combination 'off the shelf' or at short notice. Use Figure 7 to help with the planning process.

Health status of planting material

Virus-related diseases are found in both own-rooted vines and in rootstocks. When two vines – the rootstock and the scion – with different levels and types of diseases are grafted, the combination of diseases can lead to loss of vigour, yield and even vine death. Traditionally, there have been reports of incompatibility between certain scion and rootstock combinations. It now appears that the virus load carried by the scion and/or rootstock is the primary cause of incompatibility problems – particularly

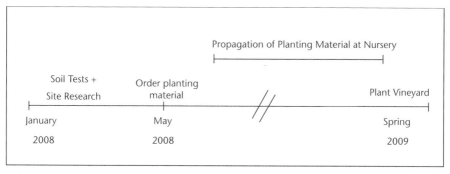

Figure 7. Recommended timeline for purchasing grafted planting material (rootstocks)

when the correct procedures and protocols are not carried through in the nursery. Lower virus loads in Australia compared with overseas has meant that the prevalence of stock-scion incompatibility is lower than in other countries. The most common problem is with Ramsey and Muscat Gordo and there are occasional reports of problems with 5BB Kober and Chardonnay.

> **To minimise the risk always purchase certified material with 'known' heath status from VINA or AVIA accredited nurseries.**

Trueness to type

Each block in a rootstock research project undertaken by PGIBSA was DNA tested to ensure trueness to type. While 38 blocks returned positive results; six blocks returned a result different from that expected. Three of these negative samples were explained by incorrect planting details, i.e. grower had mixed up the rootstock identity of two blocks. The other three negative results could not be explained by the grower, indicating that the material supplied was different to that ordered. These results again highlight the importance of using certified material from accredited nurseries as it is the best way to ensure that the material ordered is the material that arrives.

Managing Rootstocks

Different rootstocks perform differently and so need to be individually managed to achieve consistent quality outcomes. Just as it would be inappropriate to manage Shiraz and Chardonnay identically, it would be inappropriate to manage two rootstocks in the same manner. The following section provides rootstock specific management advice.

Managing rootstocks according to their water requirements

Different rootstocks (and scion varieties) have different water requirements. The difference between rootstocks appears to be related to the influence of the rootstock on the development of the root system (Soar et al., 2006). Rootstock influences the size of the root system (vertical depth and horizontal spread), the root density at different depths and the proportion of small, medium and large diameter roots. Rootstock also alters other physiological mechanisms which contribute to a vine's water requirements including leaf transpiration rate (stomatal conductance), photosynthesis and hydraulic conductivity.

To ensure optimum irrigation efficiency, quality and yield it is essential that growers understand the water requirements and appropriate management strategies for the rootstocks on their vineyard. The information below is based on the literature (see footnote, page 32) and anecdotal evidence from vineyard managers and research by PGIBSA.

Group A (Schwarzmann and 101–14)[5]

These rootstocks generally have shallow, laterally spreading, low density root systems and, as a result, have less access to soil moisture and therefore greater water requirements than other rootstocks. 101–14 and Schwarzmann are both consistently classified in the literature as being susceptible to drought conditions. Research suggests that:

- 101–14 has equal or greater water requirements than ungrafted vines; and

- Schwarzmann has similar or greater water requirements than ungrafted vines but lower water requirements than 101–14.

---

5. 3309C has not been included as there is little information on managing this rootstock in South Australia

To optimise irrigation efficiency, quality and yield, growers should consider reducing dripper spacing and applying shorter, more frequent irrigations to ensure that the water is targeted at the majority of the root system. This is particularly important on coarse-textured soils where the wetting pattern spreads vertically rather than horizontally (see page 25). Any cultural practices that would help to retain moisture such as mulching would be beneficial in blocks grafted to these rootstocks and may provide the additional 'buffer' necessary to keep canopies fresh in the peak irrigation period of the season or during heat waves.

Group B (5C Teleki, 5BB Kober and SO4)[6]

These rootstocks should be managed according to their vigour, because while they generally impart moderate vigour to the scion and therefore have moderate drought tolerance, in deep soils they can impart high vigour, have more extensive root systems and therefore have higher drought tolerance. Research suggests that 5BB Kober tolerates drought conditions more readily than 5C Teleki and SO4. In general, Group B rootstocks have:

- Lower water requirements compared with the group A rootstocks (101–14 and Schwarzmann)

- Similar water requirements to ungrafted vines

Growers need to be aware, however, that Group B rootstocks are prone to strong vegetative growth early in the season and late season leaf loss. Growers should therefore apply water late rather than early in the season to minimise these rootstock effects. Mulching, particularly on coarse textured soils, would also help to retain moisture late in the season and reduce leaf loss.

Group C (1103 Paulsen, 140 Ruggeri, 110 Richter and 99 Richter)

These rootstocks have been consistently classified in the literature as being drought-tolerant and, with the right management strategy, offer the potential for water savings throughout the growing season.

These rootstocks develop large, dense root systems and, in general, have greater access to soil moisture than ungrafted vines, i.e. the rootstocks have higher levels of readily available water (RAW). To ensure optimum irrigation efficiency and quality

---

6. 420A has not been included as there is little information on managing this rootstock in South Australia

these rootstocks therefore need to be managed differently to ungrafted vines. In practice, blocks grafted to these rootstocks should be irrigated *less frequently* than ungrafted blocks. Growers who manage these rootstocks with the same irrigation schedule as ungrafted vines will find that the rootstocks produce excessive vigour with subsequent negative effects on quality. Research suggests that:

- 110 Richter has greater drought tolerance than 99 Richter and group A and B rootstocks

- 110 Richter has less drought tolerance than 140 Ruggeri and 1103 Paulsen

- 140 Ruggeri and 1103 Paulsen have similar levels of drought tolerance

Growers need to be aware that the production of excessive vigour by these rootstocks in 'wet' seasons can be an issue. The best way to overcome this is to plant a high water-use cover-crop. This will have two outcomes; firstly it will help to 'dry out' the top layers of the soil profile making the vines 'work harder' resulting in a reduction in shoot vigour; secondly it will give growers control over the water inputs to the vine earlier in the season. Growers should keep the cover-crop growing for as long as is necessary to achieve the desired outcomes. In practice, this will mean that the cover-crop in blocks grafted to these drought-tolerant rootstocks will need to be left in for longer than would be the case for ungrafted or lower vigour rootstocks.

Penfold (2006) found that chicory and most varieties of ryegrass and fescue are the most effective cover-crop species for reducing water (and nutrient) availability and decreasing vine vigour, as these varieties continue to grow strongly after slashing, so long as there is moisture available. Note: The risk of frost damage is increased with the use of a cover-crop compared with a bare soil.

Group D (Ramsey)

Ramsey is the most drought-tolerant rootstock available in Australia and has the lowest water requirements of all rootstocks. To ensure optimum irrigation efficiency and quality, this rootstock therefore needs to be managed differently. In practice, blocks grafted to Ramsey should be irrigated *less frequently* compared with other rootstocks and own roots. Growers who manage Ramsey with the same irrigation schedule as other rootstocks/own roots will find that scion varieties grafted to Ramsey will produce excessive vigour with subsequent negative effects on quality.

As with the group C rootstocks production of excessive vigour by Ramsey in 'wet' seasons can be an issue, so the same cover-crop strategy should be employed.

In summary:

- Different rootstocks have different water requirements and therefore need to be managed differently:

- Separate rootstocks (and own roots) into different irrigation shifts

- Separate moisture monitoring by rootstock

Nitrogen: influence on rootstock performance and the implications for selection and management

Nitrogen Requirements: Rootstocks vs Own roots

Nitrogen application and vine nitrogen status influences vine performance and quality. Growers need to be aware that most rootstocks take up and assimilate nitrogen more efficiently than own-rooted vines and therefore rootstocks require lower nitrogen inputs than ungrafted vines. This is particularly the case for the moderate to high vigour rootstocks such as 5BB Kober, 110 Richter, 99 Richter, 1103 Paulsen, 140 Ruggeri and Ramsey.

Lower vigour rootstocks such as 101–14, Schwarzmann and 5C Teleki may have similar nitrogen requirements as ungrafted vines.

N supply as a management option to control shoot vigour

Research by PGIBSA found that as nitrogen at flowering increased so to did pruning weights (Figure 8). These findings agree with Zerihun and Treeby (2002) who observed

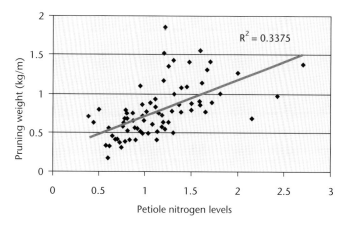

Figure 8. The influence of %N at flowering on pruning weights in 2004–2006

that biomass accumulation (in Cabernet Sauvignon) is highly responsive to nitrogen supply on all rootstocks and that nitrogen supply may be used as management option to control vine vigour.

The optimum nitrogen concentration range for grapevine petioles sampled at flowering is 0.8–1.1% (Robinson et al., 1997). In research by PGIBSA; when end-use quality[7] was compared with % nitrogen at flowering, it was observed that 73% of blocks with % nitrogen levels below 1.1% were assessed as premium or above (24% super premium and 38% premium), whereas only 27% of blocks were assessed as premium or above when % nitrogen levels increased over 1.1%. This indicates that where growers, regardless of the rootstock, scion and region, keep % nitrogen levels at flowering below 1.1% they will increase the likelihood of producing premium (or better) wine-grapes. This observation is related to the fact that, at lower nitrogen levels, vines will produce lower shoot vigour.

Rootstock response to nitrogen

Research has shown that rootstocks differ in their response to nitrogen (Keller et al., 2001a). While there has been fairly extensive research in the area it is difficult to provide any strong management advice for individual rootstocks. It is possible to say that, because rootstocks respond differently, they require different management strategies (rate and timing of applications) and should be sampled separately.

Summary
- Keep % nitrogen levels at flowering below 1.1% to increase likelihood of producing balanced vines and quality wine-grapes
- Rootstocks generally take up and assimilate nitrogen more efficiently than own-rooted vines and therefore require lower nitrogen inputs
- Rootstocks differ in their response to nitrogen application
- It is better to lean towards lower rather than higher nitrogen inputs—it is easy to apply more nitrogen but impossible to remove
- Take petiole samples at flowering to track % nitrogen levels. Aim for less than 1.1%.

---

7 Based on commercial classification

Further Reading

May, P. (1994) *Using Grapevine Rootstocks: The Australian Perspective.* (Winetitles: Adelaide).

Wolpert, J.A., Walker, M.A. and Weber, E. (1992) *Rootstock Seminar: A Worldwide Perspective.* (American Society for Enology and Viticulture: Davis).

Whiting, J (2004). Rootstocks. In *Viticulture Volume 1: Resources.* Eds. P.R. Dry, and B.G. Coombe, (Winetitles: Adelaide).

Whiting, J. (2003) *Selection of Grapevine Rootstocks and Clones for Greater Victoria.* (Department of Primary Industries: Victoria).

References

Becker, H. (1988) Börner: The first rootstock immune to all phylloxera biotypes. In *Proceedings of the Second International Cool Climate Viticulture and Oenology Symposium.* Eds R.E. Smart et al. New Zealand Society for Viticulture and Oenology, Auckland, New Zealand, p. 51.

Becker A. and Wheeler P. (2000) Suitability of *Vitis* rootstock cultivars for a Franconian 'Muschelkalk' location. *Acta Horticulturae ISHS 528: VII International Symposium on Grapevine Genetics and Breeding.*

Bramley, R.G. and Hamilton, R.P. (2005) Hitting the zone: making viticulture more precise. In Proceedings of the 12th Australian Wine Industry Technical Conference, 2004. Eds. R.J. Blair, P.J. Williams, I.S. and Pretorius (Australian Wine Industry Technical Conference), pp. 57–61.

Bramley, R.G. (2005) Understanding variability in winegrape production systems 2. Within vineyard variation in quality over several seasons. A*ustralian Journal of Grape and Wine Research* **11**, 33–43.

Carbonneau, A. (1985) The early selection of grapevine resistant rootstocks for resistance to drought conditions. *American Journal of Enology and Viticulture* **36**, 195–198.

Cirami, R., Furkaliev, J. and Radford, R. (1994) Summer drought and vine rootstocks. *Australian and New Zealand Grape grower and Winemaker,* No. 366a, 145.

Candolfi-Vasconcelos, M.C. (1995) *Phylloxera resistant rootstocks for grapevines.* North West Berry and Grape Information Network.

Cirami, R. (1999) *Guide to the selection of phylloxera resistant rootstock.* (Phylloxera and Grape Industry Board of South Australia).

Delas, J.J. et al (1991) in Delas, J.J. (1992) Criteria used for rootstock selection in France. In *Proceedings Rootstock Seminar: A Worldwide Perspective*, Reno, 1992. Eds. J.A. Wolpert, M.A. Walker et al. (American Society for Enology and Viticulture), pp 1–14.

Dry, P.R., Maschmedt, D.J., Anderson, C.J., Riley, E., Bell, S-J. and Goodchild, W.S. (2004) The grapegrowing regions of Australia. In *Viticulture Volume 1: Resources.* Eds P.R. Dry and B.G. Coombe (Winetitles: Adelaide).

Dry, P.R., Iland, P.G. and Ristic, R. (2005) What is vine balance? In *Proceedings of the 12th Australian Wine Industry Technical Conference*, 2004. Eds. R.J. Blair, P.J. Williams, I.S. and Pretorius (Australian Wine Industry Technical Conference), pp. 68–74.

Ewart, A., Gawel, R., Thistlewood, W.P., McCarthy, M. (1993) Effect of Rootstock on the composition and quality of wines from the scion Chardonnay. *Australian and New Zealand Wine Industry Journal* **8** (3), 270–274.

Ezzahouani, A. and Williams, L. (1995). The influence of rootstock on leaf water potential, yield, and berry composition of ruby seedless grapevines. *American Journal of Enology and Viticulture*. **46**: 559–563.

Galet, P. (1998) *Grape Varieties and Rootstock Varieties*. (Oenoplurimédia), France

Gibberd, M., Walker, R., Blackmore, D. and Congdon, A. (2001) Transpiration efficiency and carbon isotope discrimination of grapevines grown under well watered conditions in either glass house or vineyard. *Australian Journal of Grape and Wine Research* 7, 110–117.

Gawel, R., Ewart, A.J.W., Cirami, R. (2000) Effect of Rootstock on the composition aroma and flavour from the scion Cabernet Sauvignon grown at Langhorne Creek. *Australian and New Zealand Wine Industry Journal*, **15** (1), 67–73.

Guillon, J. M. (1905) *Étude générale de la vigne: Historique les vignobles et les crus anatomie et physiologie, sol et climat*. Masson, Paris.

Gladstones, J. (1992) *Viticulture and Environment*. (Winetitles: Adelaide).

Henschke, P. (2007) The influence of three rootstocks on vine performance, composition and the wine sensory properties of early, mid and late maturing clones of Pinot Noir. *Australian and New Zealand Grapegrower and Winemaker* 512a, 38–40.

Keller, M., Kummer, M. and Candolfi-Vasconcelos, M. (2001a) Soil nitrogen utilisation for growth and gas exchange by grapevines in response to nitrogen supply and rootstock. *Australian Journal of Grape and Wine Research* 7, 2–11.

Krstic, M. and Hannah, R. (2003) Matching Scion and Rootstock combinations in Sunraysia. Final Report to GWRDC. Project No. RT02/19–3 (Department of Primary Industries: Mildura).

Maschmedt, D. (2004). Soils. In *Viticulture Volume 1: Resources*. Eds. P.R. Dry and B.G. Coombe (Winetitles: Adelaide).

Maschmedt, D., Nicholas, P.R. Cass, A. and Myburgh, P.A. (2004) Soil physical properties. In *Soil, Irrigation and Nutrition*. Ed. P.R. Nicholas (Winetitles: Adelaide).

May, P. (2004) *Flowering and Fruitset in Grapevines*. (Lythrum Press: Adelaide).

McArthy, M., Cirami, R., Furkaliev, J. (1997) Rootstock response to Shiraz to dry and drip irrigated conditions. *Australian Journal of Grape and Wine Research* **3**, 95–98.

Nicol, J.M., Stirling, G.R., Rose, B.J., May, P. and Van Heeswijck, R.V. (1999) Impact of Nematodes on grapevine growth and productivity: Current knowledge and future directions, with special reference to Australian viticulture. *Australian Journal of Grape and Wine Research* 5, 109–127.

Pech, J., Stevens, R.M. and Gibberd, M.R. (2002) Responses of Chardonnay and Shiraz on five rootstocks to reduced irrigation. Poster Summary. Proceedings of the 11th Australian Wine Industry Technical Conference. Australian Wine Industry Technical Conference, Adelaide. p. 222.

PIRSA Land Information (2000) *Assessing Agricultural Land*. [CD ROM]. Primary Industries and Resources South Australia.

Porten, M., Schmid, J. and Rühl, E. Current problems with phylloxera on grafted vines in Germany and ways to fight them. In *Proceedings of the International Symposium: Grapevine Phylloxera Management*. Eds K.S. Powell and J. Whiting, Department of Natural Resources and Environment, Victoria, pp. 89–97.

Quader, M., Riley, I. and Walker, G. (2002) Nematode parasites in South Australian vineyards. *Australian and New Zealand Grapegrower and Winemaker*. No. 464, 62–64.

Schmid, J., Sopp, E., and Rühl, E. (1998) Breeding rootstock varieties with complete phylloxera resistance. *Acta Horticulturae* 473

Smart, D.R., Schwass, E., Lakso, A. and Morano, L. (2006) Grapevine rooting patterns: a comprehensive review. *American Journal of Enology and Viticulture* 57 (1), 89–101.

Soar, C.J., Dry, P. R. and Loveys, B. R. (2006) Scion photosynthesis and leaf gas exchange in *Vitis vinifera* L. cv. Shiraz: Mediation of rootstock effects via xylem sap ABA. *Australian Journal of Grape and Wine Research* 12, 82–96.

Southey, J.M. (1992) Grapevine rootstock performance under diverse conditions in South Africa. In *Proceedings Rootstock Seminar: A Worldwide Perspective*, Reno, 1992. Eds. J.A. Wolpert, M.A. Walker et al. (American Society for Enology and Viticulture), pp. 27–37.

Stirling, G.R., Stanton, J.M. and Marshall, J. (1992) The importance of plant-parasitic nematodes to Australian and New Zealand agriculture. *Australasian Plant Pathology* 24, 104–115.

Swanpoel, J.J. and Southey, J.M. (1989) The influence of rootstock on the rooting pattern of the grapevine. *South African Journal of Enology and Viticulture* 10 (1), 23–27.

Tee, E. and Burrows, D. (2004) *Best irrigation management practices for viticulture in the Murray Darling Basin.* (Cooperative Research Centre for Viticulture: Glen Osmond).

Virgona, J.M., Smith, J., Holzapfel, B. (2003) Scions influence apparent transpiration efficiency of *Vitis Vinifera* (cv. Shiraz) rather than rootstocks. *Australian Journal of Grape and Wine Research* 9, 183–185.

Vrišič, S., Valdhuber, J., and Pulko, B. (2004) Compatibility of the rootstock Börner with various scion varieties. *Vitis* 43(2), 155–156.

Walker, G. (2002) Lesion Nematodes: Identification of Grapevine Resistance. Final Report for Project Number: SAR 99/5 to GWRDC.

Walker, R. and Stevens, R. (2004). Recent developments in the understanding of the effects of salinity on grapevines. Unpublished.

Walker, R., Blackmore, D., Clingeleffer, Godden, P., Francis, L. Valente, P. and Robinson, E. (2002) The effects of salinity on vines and wines. *Australian Viticulture* 6 (4), 11–21.

Walker, R.R. et al. (1998) Effects of the rootstock Ramsey (*Vitis champini*) on ion and organic acid composition of grapes and wine, and on wine spectral characteristics. *Australian Journal of Grape and Wine Research* 4, 100–110.

Walker, R.R., Read, P.E. and Blackmore, D.H. (2000) Rootstock and salinity effects on rates of berry maturation, ion accumulation and colour development in Shiraz grapes. *Australian Journal of Grape and Wine Research* 6, 227–239.

Walker, R. (2004). Application of carbon isotope discrimination technology to understanding and managing wine grape water use efficiency. CRCV Final Report to GWRDC. Project Number CRCV 99/10.

Whiting, J. and Orr, K. (1990) 99 Richter and 101–14 rootstocks susceptible to waterlogging. *Australian Grapegrower & Winemaker* No. 321, 60.

Whiting, J. (2003) *Selection of Grapevine Rootstocks and Clones for Greater Victoria.* (Department of Primary Industries: Victoria).

Zhang, X., Walker, R.R., Stevens, R.R., and Prior, L.D. (2002) Yield-salinity relationships of different grapevines (*Vitis Vinifera* L.) scion-rootstock combinations. *Australian Journal of Grape and Wine Research* 8, 150–156.

Appendix 1

Rootstock Characteristics Summary

Group	Rootstock	Relative Vigour	Nematode Resistance (Root-Knot)	Lime Tolerance	Salinity Tolerance	Drought Tolerance	Soil Acidity	Water-logging
A	101-14	L	H	▶	MT	S	○	○
	Schwarzmann	L-M	H	▶	MT	S	○	▶
	3309C	L	L	▶	S	S	○	○
B	SO4	M	M-H	●	MS	MS	○	▶
	5C Teleki	M	M	●	MS	MS	○	○
	5BB Kober	M-H	H	●	MS	MS	○	○
	420A	L	M	●	S	S	○	○
C	110 Richter	M	M	●	MS	T	▶	▶
	1103 Paulsen	M-H	M-H	●	MT	T	▶	▶
	99 Richter	M-H	M-H	●	MT	T	▶	○
	140 Ruggeri	M-H	H	●	MT	HT	●	○
D	Ramsey	H	H	▶	MT	HT	○	○
E	*Vitis Vinifera*	M-L	○	●	S	MS	▶	▶

Key to table:

● Good   ▶ Moderate   ○ Poor   L = Low   M = Medium   H = High
S = Susceptible   MS = Moderately susceptible   MT = Moderately tolerant
T = Tolerant   HT = Highly Tolerant

Notes for the table

Groups	A = *V. riparia x V. rupestris*
	B = *V. berlandieri x V. riparia*
	C = *V. berlandieri x V. rupestris*
	D = *V. champini*
	E = *V. vinifera (included for comparison)*
Rootstock	All rootstocks are phylloxera resistant
	Older 5A Teleki identified as 5BB Kober
Vigour	Refers to the relative vigour conferred to the scion. This will also be dependent on site and management.
Nematodes	Rootstock resistance to root-knot nematode
Lime	Ability of rootstock to tolerate varying degrees of active lime
Acid Soil	Performance in acidic soils
Salinity	Ability of rootstock to tolerate varying degrees of soil salinity
Drought Tolerance	Indicates tolerance of rootstock to drought (limited soil moisture) conditions during the growing season
Waterlogging	Refers to performance of rootstocks in soils that are waterlogged during the growing season or periods of root growth.

This table was adapted from a table originally developed by Phil Nicholas, SARDI.

Appendix 2

Planting Statistics

GI	Total Planted Area (Ha)			Rootstock Area (Ha)			% Rootstock		
	Total as of 2006	1997-2006	2002-2006	Total as of 2006	1997-2006	2002-2006	As of 2006	1997-2006	2002-2006
Adelaide Hills	3,810.1	2,917.2	773.9	172.3	153.8	55.4	4.5	5.3	7.2
Adelaide Plains	591.6	273.7	77.5	149.4	143.5	63.6	25.2	52.4	82.1
Barossa Valley	10,334.4	5,367.6	1,813.2	1,971.9	1,234.0	498.9	19.1	23.0	27.5
Barossa Zone	105.5	105.5	2.2	–	–	–	–	–	–
Bordertown	1,273.2	968.1	71.4	73.3	73.3	–	5.8	7.6	–
Clare Valley	5,714.4	3,329.3	708.6	144.6	115.6	45.1	2.5	3.5	6.4
Coonawarra	5,794.4	2,444.9	554.0	85.0	65.8	40.0	1.5	2.7	7.2
Currency Creek	913.9	820.2	248.5	50.5	50.5	29.7	5.5	6.2	12.0
Eden Valley	2,226.8	1,021.2	225.7	292.9	196.6	94.3	13.2	19.3	41.8
Far North Zone	4.0	4.0	–	–	–	–	–	–	–
Fleurieu Zone	218.8	174.0	98.2	1.0	1.0	1.0	0.5	0.6	1.0
Kangaroo Island	112.6	85.2	12.3	3.1	1.9	1.9	2.8	2.2	15.4
Langhorne Creek	6,251.3	4,079.1	1,043.9	934.0	707.2	346.1	14.9	17.3	33.2
Limestone Coast Zone	522.1	290.7	49.4	7.4	7.4		1.4	2.5	–
Lower Murray Zone	444.1	314.6	78.2	201.8	154.1	64.5	45.4	49.0	82.6
McLaren Vale	7,005.4	3,266.0	583.1	542.5	346.4	117.1	7.7	10.6	20.1
Mount Benson	488.9	388.4	99.9	137.2	136.5	29.1	28.1	35.1	29.1
Mount Lofty Zone	285.0	239.6	23.3	1.7	1.5	1.4	0.6	0.6	6.1
Padthaway	4,036.7	1,613.4	573.5	329.1	230.5	166.6	8.2	14.3	29.1
Riverland	22,112.9	11,938.3	3,338.8	8,886.8	5,689.3	2,524.3	40.2	47.7	75.6
Southern Fleurieu	409.8	339.0	61.1	51.3	50.8	11.4	12.5	15.0	18.6
Southern Flinders Zone	166.8	152.7	24.2	1.2	1.2	–	0.7	0.8	–
The Penninsulas Zone	35.2	27.4	10.4	10.3	10.3	2.3	29.2	37.5	21.7
Wrattonbully	2,560.5	1,866.2	307.3	206.3	168.8	80.9	8.1	9.0	26.3
Total	75,418.2	42,026.0	10,778.5	14,253.2	9,539.7	4,173.6	18.9	22.7	38.7

Appendix 3

On-farm Rootstock Trials: the why and how

There are a number of reasons why it is a good idea to set up an on-farm rootstock trial:

> ➤ Pre-planning to find out which rootstocks are most suitable for your site, and how best to manage them. That way, if phylloxera does ever appear in your region, you will be well prepared to make the best replanting decision for your specific site considerations.

> ➤ Depending on the size of your trial, you would also have at least a small area of vines that would continue to produce a full crop in the event of your vineyard being infested with phylloxera, giving you a secure income stream while you restructured the remaining area.

> ➤ If you have (or may have in the future) a particular site issue such as salinity or reduction in water availability, that may be improved by using rootstocks, it is best to conduct a trial to see which rootstock performs best in your situation, before you do a large-scale new planting.

> ➤ For scientific experimentation, either conducted by the grower or a researcher.

When setting up a trial attempt to minimise any sources of possible variation so that any differences observed in your trial are due to the different rootstocks and not variations in soil, micro-climate or any other environmental factor.

There are ways to get around variations within your block - so long as you know where the variation exists. The following designs are easy to set up and maintain and will still provide good information on the differences between rootstocks. Remember: only make the trial as large as you can handle. There is no point in having a large trial if you don't have the time to collect and analyse the data.

Design No. 1

With a limited area in which to conduct a trial (3–9 rows) and a soil change in one direction, it is possible to set up a trial along the lines of the one in Figure 9. In this situation the change in soil type will have an equivalent effect on all three rootstocks thereby removing any source of possible variation.

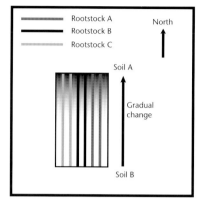

Figure 9. Rootstock trial design for a site with limited area available and with gradual soil change down the row.

Design No. 2

If there is more area available for a trial (10–20 rows) a randomised block design with whole row plots may be used. In Figure 10, the soil type changes from north to south and there is a gradual slope running from the north-east corner of the block to the south-west corner, creating different environmental conditions across the block. The design below will account for both soil and environmental variations as long as you randomly assign the rows and your sampling technique is sound.

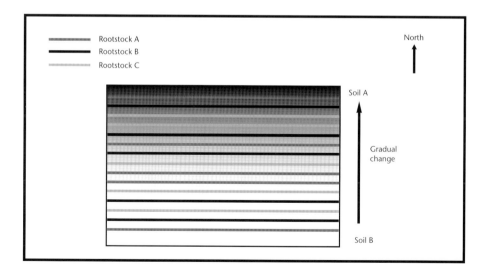

Figure 10. Rootstock trial design for a site with a gradual soil change across the rows.

Design No. 3

Where a grower has a large amount of room to devote to a trial (20+ rows) and the soil is uniform or changes in one direction, it is possible to set up each block of rootstocks in management units with different irrigation shifts (Figure 11). This will allow the grower to manage each rootstock to its optimum and measure performance in terms of income (price per tonne × yield) – costs (inputs). The diagram below gives an example of how it could be set up.

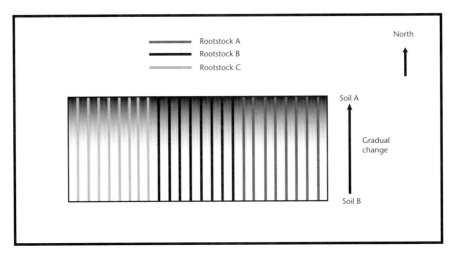

Figure 11. Rootstock trial design for a site with a large area available and with gradual soil change down the row.

Which rootstocks should I plant?

The choice of a shortlist of rootstocks to compare depends on your soil type, the scion variety and other factors such as whether or not you have nematodes. Read the preceding chapters in this guide, talk to your local vine improvement association, or nursery to get advice on which rootstocks to plant.

Which data should I collect?

Generally, most growers are after an indication of vine performance (yield and quality). There are a number of ways to measure yield, however quality can be a little more difficult to determine. Small batch wines can be made at Provisor (tel: (08) 8303 7359) or you can simply ask your company winemaker to duck into your trial block and try the fruit. Red grapes can be measured for colour (mg/g anthocyanins), white grapes can undergo a G-G assay, visit the Australian Wine Research Institute

(AWRI) website for more information **www.awri.com.au**

Other measurements that may be of interest are vine balance (yield to pruning weight ratio); yield components (berry and bunch number and weight) and grape composition (Brix, pH and TA) leading up to harvest. Remember the key to getting meaningful results is taking appropriate samples.

Sampling

The collection of *representative* samples is very important. If samples are not representative of all the vines on a given rootstock, then your conclusions may well be wrong. Most growers know how to sample their vineyard for yield estimation, pre-vintage quality assessments or problem diagnosis. If you would like more information, contact your grower liaison officer, AWRI or the Board.

For more guidance on establishing a trial or planting on rootstocks, contact the Rootstock Project Officer at the Board's office – (08) 8362 0488.

LYTHRUM PRESS
PO Box 243 Rundle Mall, Adelaide, South Australia 5000

www.lythrumpress.com.au